Nicola Schoenenberger

Gene flow from wheat (Triticum aestivum L.) to Aegilops L. species

Nicola Schoenenberger

Gene flow from wheat (Triticum aestivum L.) to Aegilops L. species

Elements for ecological risk assessment of genetically modified crops

Südwestdeutscher Verlag für Hochschulschriften

Impressum/Imprint (nur für Deutschland/ only for Germany)
Bibliografische Information der Deutschen Nationalbibliothek: Die Deutsche Nationalbibliothek verzeichnet diese Publikation in der Deutschen Nationalbibliografie; detaillierte bibliografische Daten sind im Internet über http://dnb.d-nb.de abrufbar.
Alle in diesem Buch genannten Marken und Produktnamen unterliegen warenzeichen-, marken- oder patentrechtlichem Schutz bzw. sind Warenzeichen oder eingetragene Warenzeichen der jeweiligen Inhaber. Die Wiedergabe von Marken, Produktnamen, Gebrauchsnamen, Handelsnamen, Warenbezeichnungen u.s.w. in diesem Werk berechtigt auch ohne besondere Kennzeichnung nicht zu der Annahme, dass solche Namen im Sinne der Warenzeichen- und Markenschutzgesetzgebung als frei zu betrachten wären und daher von jedermann benutzt werden dürften.

Verlag: Südwestdeutscher Verlag für Hochschulschriften Aktiengesellschaft & Co. KG
Dudweiler Landstr. 99, 66123 Saarbrücken, Deutschland
Telefon +49 681 37 20 271-1, Telefax +49 681 37 20 271-0, Email: info@svh-verlag.de
Zugl.: University Neuchâtel, Faculty of Science, Neuchâtel, Switzerland, PhD thesis, 2005

Herstellung in Deutschland:
Schaltungsdienst Lange o.H.G., Berlin
Books on Demand GmbH, Norderstedt
Reha GmbH, Saarbrücken
Amazon Distribution GmbH, Leipzig
ISBN: 978-3-8381-0510-9

Imprint (only for USA, GB)
Bibliographic information published by the Deutsche Nationalbibliothek: The Deutsche Nationalbibliothek lists this publication in the Deutsche Nationalbibliografie; detailed bibliographic data are available in the Internet at http://dnb.d-nb.de.
Any brand names and product names mentioned in this book are subject to trademark, brand or patent protection and are trademarks or registered trademarks of their respective holders. The use of brand names, product names, common names, trade names, product descriptions etc. even without a particular marking in this works is in no way to be construed to mean that such names may be regarded as unrestricted in respect of trademark and brand protection legislation and could thus be used by anyone.

Publisher:
Südwestdeutscher Verlag für Hochschulschriften Aktiengesellschaft & Co. KG
Dudweiler Landstr. 99, 66123 Saarbrücken, Germany
Phone +49 681 37 20 271-1, Fax +49 681 37 20 271-0, Email: info@svh-verlag.de

Copyright © 2009 by the author and Südwestdeutscher Verlag für Hochschulschriften Aktiengesellschaft & Co. KG and licensors
All rights reserved. Saarbrücken 2009

Printed in the U.S.A.
Printed in the U.K. by (see last page)
ISBN: 978-3-8381-0510-9

ABSTRACT

Hybridisation and introgression from crops to wild relatives is a key issue in risk assessment. In the present study, hybridisation and introgression dynamics from hexaploid wheat (2n=42) to tetraploid *Aegilops* species (2n=28) were investigated by experiments in natural conditions or in the greenhouse, and by genetic analyses.

In order to study crop-weed hybridisation as a function of distance, a field trial was set up where *Ae. cylindrica* was planted in plots at 0, 1, 5, 10 and 25m from a wheat field. In the progeny (14045 seeds sown) we detected hybrids up to 1m from the wheat field.

Wheat-specific RAPD fragments were found in *Ae. cylindrica* x *T. aestivum* hybrids and BC_1 plants. Using a set of Chinese Spring nulli-tetrasomic wheat lines, we were able to assign DNA fragments to wheat chromosomes. Introgressed wheat-specific markers were localised on the three genomes (A, B and D). Some of these markers were transformed into easy-to-use Sequence Characterised Amplified Regions (SCARs), and used to characterise an introgressive series.

Ae. cylindrica x GM-wheat hybrids, BC_1 and BC_1S_1 (self-fertilised first backcrosses) were manually produced, in order to study inheritance of transgenes. Female fertility of the hybrids was 0.03-0.6%, BC_1 plants had 30-84 chromosomes and displayed highly irregular meioses, their self fertility ranged from 0 to 5.21 %. BC_1S_1 plants had 28-43 chromosomes and some of them recovered full fertility. One BC_1S_1 individual contained the *bar* gene issued from its transgenic wheat progenitor and survived herbicide treatment.

A RAPD-based population genetics study was carried out in natural *Ae. cylindrica* populations, most of them from adventive locations in Switzerland, Italy and the USA. Genetic diversity was low and most of the variance resided among populations. Italian populations from the Aosta valley and a Swiss population were similar or identical to northern American populations indicating that the species may have crossed the Atlantic Ocean several times.

Using 52 *Ae. geniculata* individuals planted in a winter wheat field, we obtained an overall *Ae. geniculata* x *T. aestivum* hybridisation rate of 0.94%, hybrid fertility, i.e. BC_1 production rate, was 2.2%. Specific wheat SCAR markers were detected in the hybrids and backcrosses.

Gene flow is thus possible between wheat and *Aegilops*, at a limited rate. Because of the large cultivation of wheat, trans-gene flow might occur significantly. Moreover, insertion of the transgene on A and B genome does not prevent introgression.

TABLE OF CONTENTS

Preface 5

Introduction 6

Summary of the chapters 8

Conclusion 12

Chapter 1:
 Gene flow by pollen from wheat to jointed goatgrass (*Aegilops cylindrica*)
 as a function of distance: a field experiment 13

Chapter 2:
 Introgression of wheat DNA markers from A, B and D genomes in early
 generation progeny of *Aegilops cylindrica* Host *Triticum aestivum* L. hybrids 28

Chapter 3:
 Molecular analysis, cytogenetics and fertility of introgression lines
 from transgenic wheat to *Aegilops cylindrica* Host 47

Chapter 4:
 RAPDs reveal low genetic variability in natural *Aegilops cylindrica* Host
 populations in Europe and Northern America 71

Chapter 5:
 Gene flow from Wheat (*Triticum aestivum* L.) to *Aegilops geniculata* Roth. 89

"The sexual transfer of genes to weedy species to create a more persistent weed is probably the greatest environmental risk of planting a new variety of crop species"[1]

PREFACE

In 1993, François Felber started a research project on gene flow from cultivated plants to related wild species. The aim was to conduct risk assessment of the commercial cultivation of genetically modified organisms (GMOs). The first investigated species were alfalfa, barley, and oilseed rape. The research was aimed at finding answers at a Swiss scale. Dessislava Savova-Bianchi and Julia Keller-Senften led the story to a success, and it was continued. This time the most widely grown crop was chosen, wheat (*Triticum aestivum* L.), which has an extremely rare and good looking wild relative growing in Switzerland, *Aegilops cylindrica* Host, alias Jointed Goatgrass. This species which is delighting any passionate Swiss botanist, is at the same time a horror to many Northern American wheat farmers and agronomists, as it is one of the major agricultural weeds there. Superfluous to say that the impact of the research conducted by the team of François Felber became more global. The first results on wheat, produced by Roberto Guadagnuolo, were quite impressive. DNA from the crop, which is a hexaploid with 42 chromosomes, could be introgressed naturally into tetraploid *Ae. cylindrica* in only two generations. *Ae. cylindrica* x *T. aestivum* hybrids were produced at frequencies of 1 to 7% and first backcrosses having *Ae. cylindrica* as the recurrent parent had euploid (2n=28) or one supernumerary chromosome. The morphology of these fertile plants was like the one of their wild progenitor but they still contained wheat-specific DNA.

The third cycle of PhD programs in the gene flow group of Neuchâtel started at this stage. Luigi D'Andrea started to work on lettuce, and I carried on the wheat subject. There where many ideas in the air, some of them of immediate importance others which would lead to results on the long term.

---

[1] Goodman RM, and Newell N. 1985. Genetic engineering of plants for herbicide resistance: status and prospects. In: Halvorson HO, Pramer D, and Rogul M. Engineered organisms in the environment: scientific issues. Pages 47-53. American Society for Microbiology, Washington DC.

INTRODUCTION

Current public controversy over the use of genetically modified (GM) crops has escalated and has led to a temporary ban in most European countries. A main concern in biosafety is the possibility of transgene escape into wild relatives through sexual gene flow. Expression of a transgenic character in wild plants could result in increased performance and weediness. Moreover, those genes could be selected by cultural practices in agroecosystems (in the case of herbicide resistance), leading to the impossibility of weed control. Very little is known on possible pleiotropic effects in transgenic hybrids and subsequent backcross generations. It has been demonstrated that gene flow may occur in nature and that it depends upon the taxon and the region investigated. Understanding the mechanisms of gene transfer between crops and wild relatives is of high importance for risk analysis of transgenic plants. These mechanisms are particularly complicated in the wheat - *Aegilops* group, a species complex with a reticulate evolutionary history, where hybridisation, polyploidy and domestication have played a major role in its evolution. The wheat - *Aegilops* group contains wild diploids and tetraploids, and cultivated tetraploids and hexaploids. Several diploid gene pools, which evolved from a common ancestor, participated in the formation of the allopolyploids. The diploid genomes maintained different levels of homeology between each other which allows gene flow to occur among them.

The aim of the present thesis was to get a better understanding of the mechanisms of introgression between wheat and its wild relatives, on one hand its ecological aspects and on the other the genetic components. We focussed on the species which may be most exposed to gene flow from new wheat cultivars, these are *Ae. cylindrica* because of its presence as an infesting weed in wheat fields, and *Ae. geniculata* because of its wide distribution around the Mediterranean and its known ability to hybridise with wheat. Hybridisation frequencies and fertilities are fundamental data needed for risk assessment of transgenic plants. On the other hand, we needed a better knowledge of the genetic mechanisms of introgression, comprising chromosome constitution, molecular marker inheritance and transgene expression in introgressive series. Moreover, genetic variability which may be a factor influencing sexual compatibility between species is a highly investigated subject in wheat, whereas little was known about it in *Ae. cylindrica*, the main wild species studied in this thesis.

Hereafter, the main ideas and results of my thesis are exposed, in the same order as the five chapters that follow this introduction. Each chapter will be the object of a scientific publication.

SUMMARY OF THE CHAPTERS

First, we wondered about isolation distances. If we start to plant transgenic wheat we want to avoid introgression of transgenes into the wild species, because they could confer a selective advantage to a population of *Ae. cylindrica* that eventually becomes more invasive. That is why we want to know how far a wheat field must be from *Ae. cylindrica* in order to avoid hybridisation. In other words, how far can a wheat pollen grain fly and still compete against *Ae. cylindrica* pollen to fecundate its ovules? We obtained an answer to that question by planting *Ae. cylindrica* plants at given distances from a wheat field, collecting the seeds produced, and counting the hybrid offspring. Wheat is a highly selfing species with pollen that stays viable only for a short time. The *Ae. cylindrica* mother plant that produced hybrid offspring at the highest distance from the fields' edge was planted one meter form the field. This indicated that isolation distances necessary to avoid hybridisation would be small.

Second, we investigated the origin and fate of wheat DNA introgressing into *Ae. cylindrica*. As said before, bread wheat is a hexaploid. It originated by hybridisation and subsequent polyploidisation of three diploid ancestors. The first cross occurred about 200'000 years ago and involved *Triticum urartu* Thüm. (A genome) and probably an unknown close relative of *Aegilops speltoides* Tausch (B genome), to produce the first polyploid wheat with an AB genome. This plant was domesticated by humans about 15'000 years ago in the Fertile Crescent. About 8'000 years ago a cross of cultivated *Triticum dicoccum* Schulb. with *Aegilops tauschii* Coss. resulted in the formation of modern hexaploid bread wheat *Triticum aestivum* (ABD genome). By the same mechanism *Aegilops tauschii* and *Aegilops caudata* L. (C genome) combined their genomes to produce wild *Ae. cylindrica* which has a CD genome. The close evolutionary link between wheat and *Ae. cylindrica* is their common D genome ancestor. In fact, in pentaploid *Ae. cylindrica* x *T. aestivum* hybrids (ABCDD genome) the D genome chromosomes pair normally at metaphase, whereas the others remain univalents and tend to be eliminated. In a perspective of risk assessment of transgene introgression this led to the hypothesis that the wheat A and B genomes were secure places to put transgenes in order to avoid their flow to *Ae. cylindrica*. Our idea was to test this experimentally. We had wheat-specific random amplified polymorphic DNA (RAPD) markers introgressing into *Ae. cylindrica* and we had an unique tool for chromosome location of DNA in wheat, which are the Nulli-Tetrasomic (NT) wheat aneuploids developed by Ernest R. Sears in the late fifties of the last century.

By simply amplifying the introgressing fragments on the NT lines, we could identify the chromosomal origin of the fragments. Interestingly those markers were not only located on the D genome of wheat but also on the A and B genome. How is that possible? All genomes in the *Triticum-Aegilops* complex have a common ancestor and diverged from it about 40 million years ago. They still are quite similar among each other and are called homeologous genomes. Wheat genetic material from the A and B genome can introgress into *Ae. cylindrica* by homeologous recombination, resulting in translocated fragments, or whole chromosomes can be maintained in an *Ae. cylindrica* background as homologous pairs and be maintained along the generations. It is therefore not enough to place a transgene on the A or B genome of wheat to avoid its introgression. Moreover, we converted our RAPD fragments into highly specific sequence characterised amplified regions (SCARs), by sequencing and primer design. These markers are easy-to-use and important tools for tracing wheat DNA along the generations of introgressive series.

Third, we wanted to simulate a natural introgression of wheat transgenes into *Ae. cylindrica*. The aim was to recover transgenic *Ae. cylindrica* plants for further investigations, and at the same moment it was the opportunity to have detailed information about every individual plant of the series. We crossed manually near-isogenic and transgenic wheat lines bearing herbicide resistance, reporter, fungal disease resistance, and increased insect tolerance transgenes with *Ae. cylindrica*. The hybrids were backcrossed to *Ae. cylindrica* and the backcrosses selfed. As in Switzerland cultivation of GMOs in the field is forbidden, all plants were grown in a greenhouse. We were able to assess, at each generation, parameters which are essential to risk assessment like fertility, survival, chromosome constitution, introgression of SCARs from the different wheat genomes, and transgenes and their expression. First generation hybrids were mostly sterile and the few produced BC1 seeds germinated badly. The ones that survived had chromosome numbers ranging from 30 to 84. Self fertility in two BC1 plants was 0.16% and 5.21% respectively. In the recovered BC1S1 families a single plant was transgenic and herbicide resistant. Other BC1S1 plants had between 28 and 31 chromosomes, several of them carried SCARs specific to wheat A and D genomes. Interestingly, the fertility of these plants, which was higher under open pollination conditions than by selfing, did not necessarily correlate with even or euploid *Ae. cylindrica* chromosome number. Some individuals having supernumerary wheat chromosomes recovered full fertility.

Fourth, we were interested to see whether the genetic differentiation of natural *Ae. cylindrica* populations correlated with its ability to form hybrids with wheat. We searched for new Swiss localities and sampled known populations in the Italian Aosta Valley and the USA, and obtained seeds from a French and a Romanian population. In a previous study, the frequency of *Ae. cylindrica* x *T. aestivum* hybridisation varied in function of the geographic origin of the wild plants used and a difference in the genetic constitution was observed too. As in that study only three Swiss populations were used, we wanted to enlarge the number of *Ae. cylindrica* populations to be put in a wheat field for hybridisations. As genetic diversity of a wild plant is often much higher than the one of a crop, it may not be enough to test hybridisation frequencies with plants coming from a single place to get a clear picture.

Unfortunately, besides a huge amount of work, the study on hybridisation frequencies did not revealed significant differences between populations. However, the population genetics part, carried out with RAPDs, gave some very interesting results. Despite of being a successful invader, *Ae. cylindrica* is very little diverse, most diversity resided among populations as expected for inbreeding annual plants. Genetic differentiation of the populations did not necessarily reflect the geographic distribution. Italian populations were more similar to Northern American ones than to Swiss ones. In Switzerland we found several new localities of *Ae. cylindrica*, indicating a recent spread of the species. Particularly, one new Swiss population having a markedly weedy behaviour was genetically undifferentiated to a Californian population, where *Ae. cylindrica* is known as a weed. *Ae. cylindrica* must have crossed the Atlantic Ocean several times, probably in both directions.

Last, we performed a rather small field experiment to characterise gene flow between wheat and *Aegilops geniculata* Roth in natural conditions. *Ae. geniculata* is one of the most common species of the genus, spread all around the Mediterranean basin, present in Switzerland sporadically as a very rare adventive. Hybrids with wheat can be found in sympatric populations of the two species. Cultivation of genetically modified wheat in Southern Europe will probably have a more important impact on *Ae. geniculata* than on *Ae. cylindrica* due to its larger distribution. We planted *Ae. geniculata* plants originating from four populations in the middle of two wheat fields, monitored them to see whether they were flowering at the same time, which was the case, and collected the seeds at the end of the season. Hybridisation rate varied between 0 and 6.06% according to field locations and population origins and mean fertility of the hybrids was 2.17%. Two recovered BC1 plants had 55 and 56 chromosomes and were carrying wheat-specific SCAR markers. The fertility of *Ae. geniculata* hybrids is higher than in *Ae. cylindrica* hybrids, meaning that gene flow from wheat might occur at a higher frequency to the former species.

CONCLUSION

The major risk of cultivating transgenic wheat may be represented by gene flow to its wild relatives, particularly to *Ae. cylindrica*. The work presented here brings some new, fundamental evidence that introgression between wheat and *Ae. cylindrica* can occur. Genetic material from all three wheat genomes can be transmitted to *Ae. cylindrica* and be stabilised in the new genetic background. Transgenes can introgress and confer new properties to the wild species. However, gene flow will occur at the field level, to wild plants growing within or just at the edge of the cultivated crop. Genetic variability in *Ae. cylindrica* is low and we were unable to relate population differentiation with hybridisation frequencies, but we got an insight into the genetic structure of adventive *Ae. cylindrica* populations, and detected close relationships particularly between Italian and Northern American populations and between some recently introduced Swiss populations and a Californian one. Gene flow from wheat is possible to *Ae. geniculata* too, maybe at a higher frequency than to *Ae. cylindrica*. This result is highly important for risk assessment of growing transgenic wheat in Southern Europe, where this species is widespread and often found in wheat growing areas.

The research conducted here is mainly focussed to elucidate the exposure component of risk of growing new wheat varieties, i.e. the probability that gene flow occurs. A major outlook is to make the hazard component clear now, i.e. the consequences of transgene flow to wild relatives. Particularly interesting are aspects like the evaluation of the fitness that a particular transgene may confer to a wild population, or the search for pleiotropic effects of a transgene. Research can be conducted on the exposure component too. Questions like the possibility of crop to wild to wild gene flow are still unanswered. Can a wild relative serve as a bridge for gene flow to occur from a cultivated crop to a more distantly related wild species? We know that different genomic regions of wheat have different probabilities to introgress into *Ae. cylindrica*. It would be useful to find genomic regions which have no possibility to be introgressed into wild species. These locations would be ideal to insert transgenes in order to avoid possible environmental harm. We look forward that these two new areas of research are presently investigated in the laboratory.

CHAPTER 1

Gene flow by pollen from wheat to jointed goatgrass (*Aegilops cylindrica*) as a function of distance: a field experiment

Abstract

In 2002, two field tests were conducted in conventional winter wheat fields in Switzerland, in order to assess jointed goatgrass x wheat hybridisation frequencies as a function of distance. The first test consisted in 262 *Ae. cylindrica* plants, planted in plots within and adjacent to a wheat field of the variety 'Titlis', at distances of 0, 1, 5, 10 and 25 m from the pollen source direction South. In the second site 213 jointed goatgrass plants were planted at 1, 5, 10 and 25 m from a wheat field of the variety 'Galaxie' direction East. Relative to the main winds on the Swiss plateau, the plots were placed downwind from the pollen source, inside oat fields. Flowering period of the two species overlapped. All the seeds of jointed goatgrass were collected in the two sites and 13 hybrids were detected both morphologically and with a DNA marker among the total 14045 offspring sown. All hybrids originated from the same field test ('Titlis'), 12 hybrids were produced inside the wheat field (0 m) and one hybrid at 1 m. Hybridisation frequencies of 3.16% and 0.32% were observed for 0 m plots. At 1 m, hybridisation frequencies were 0% and 0.29%. No hybrids were detected at distances beyond 1 m. Wind speed and direction, temperature evolution, relative humidity and precipitations were registered during pollination at the field site. The results suggest that the distances required to efficiently avoid introgression from wheat to jointed goatgrass are small, of the order of a few meters.

Introduction

Crop-to-wild gene flow happens naturally in most domesticated plants when they come into contact with their wild relatives, the possible harmful consequences being the evolution of more aggressive weeds and the increased likelihood of extinction of wild taxa (Ellstrand et al. 1999). Gene flow and its consequences represent one of the main concerns over the agronomical use of Genetically Modified (GM) crops (Wolfenbarger and Phifer 2000), particularly because engineered genes for pest or disease resistance, or resistances to abiotic stress like drought, frost, salinity or herbicide use, usually confer a selective advantage to the modified crop and possibly to the wild relative receiving those genes through introgression (Ellstrand and Hoffman 1990, Ellstrand 2003).

In order to study mechanisms of gene flow from GM or conventional crops to wild relatives quantifying potential hybridisation is a key issue. Although a much more accurate evaluation of gene flow is required today to avoid seed contamination with transgenes, quantification of hybridisation is not new to plant breeders and seed producers, who have been concerned with it for a long time. In particular, studies on seed purity maintenance produced considerable results on gene flow, on pollen dispersal and on required isolation distances to minimise it (e.g. Bateman 1947). Up to now several transgenic wheat lines have been produced. Engineered traits include herbicide, pathogen and herbivore resistances; drought tolerance, male sterility and improvement of dough elasticity. Moreover, the first report of a cereal crop used for the expression of antibody molecules refers to wheat (Janakiraman et al. 2002, www.isb.vt.edu).

Ae. cylindrica is a Mediterranean/Western Asiatic species, and is present as an adventive in Western and Northern Europe and the USA (Van Slageren 1994). Although *Ae. cylindrica* is known to be a serious weed in the USA, it does not only grow as an infesting plant in wheat fields, but also in rangeland surrounding wheat growing areas, or in fence rows and roadsides outside wheat fields (Donald and Ogg 1991). In California, *Ae. cylindrica* is restricted to roadsides and pastures, whereas in Oregon it grows at the edge and inside wheat fields. Hybrids between the two species were found within or close to wheat fields (Watanabe and Kawahara 1999). Moreover, some hybrid collection sites in Oregon were not directely associated with infested wheat fields, but rather located in adjacent *Ae. cylindrica* populations (Morrison et al. 2002a). In its native range, *Ae. cylindrica* grows mainly in ruderal habitats, wastelands, road and railwaysides, dry grasslands, mountain slopes, and close or within cultivation (van Slageren 1994). In Hungary it was found to grow near wheat fields and to hybridise with the crop (Rajhathy 1960). In Western Europe and particularly in Switzerland *Ae. cylindrica* is considered a rare adventive species and is included in Switzerland's Red List of Threatened Taxa and classed in the IUCN category VU (vulnerable), furthermore it represents the only species of the genus that is constantly present in the Swiss flora (Moser et al. 2002). Among the 22 recognised *Aegilops* species, *Ae. cylindrica* is one of the most weedy ones (van Slageren 1994). For wheat, gene flow was demonstrated between hexaploid wheat and its wild tetraploid relative *Ae. cylindrica* (jointed goatgrass) (Zemetra et al. 1998; Guadagnuolo et al. 2001). The two species are close relatives (van Slageren 1994), gene flow is facilitated by the fact that the two species share the D genome. In hybrids, these chromosomes pair successfully at meiosis, resulting in recombination between D genome chromosomes of the two parental species (Zemetra et al. 1998). Both winter and spring wheat varieties overlap in phenology with *Ae. cylindrica* at least in some areas of their distribution, resulting in the possibility of hybrid formation (Guadagnuolo et al. 2001, Morrison et al. 2002a).

Hybridisation frequencies ranged from 1 to 7% in Switzerland, and 85 hybrid plants pollinated by *Ae. cylindrica* produced 13 BC1 (Guadagnuolo et al. 2001). Specific wheat RAPD fragments and one microsatellite were detected in those BC1, showing that introgression from wheat to *Ae. cylindrica* can occur naturally. First backcross individuals with partially recovered fertility, were obtained from *Ae. cylindrica* x *T. aestivum* hybrids pollinated either by *Ae. cylindrica* or by wheat in the USA (Morrison et al. 2002b). Hexaploid wheat may also hybridise with other *Triticum* and *Aegilops* species, as reviewed in Hedge and Waines (2004). Hybridisation as a function of distance involving wheat and wild relatives has never been addressed so far, while Matus-Cadiz et al. (2004) described gene flow up to 1 m distance between cultivated bread and durum wheat (*Triticum turgidum* L.). On the contrary to interspecific gene flow, intraspecific gene flow in bread wheat has led to many studies. Wheat is predominantly an autogamous species; pollen flow, if any, is due to wind (Feil and Schmid 2001). Outcrossing rates are mostly around 1 %, but can exceed 5% depending on the varieties (Martin 1990, Hucl 1996). The discovery of male sterile wheat and the consequent possibility of hybrid wheat seed production engendered several studies on flowering biology of wheat, pollen flow, pollen viability and cross fertilisation distances (Joppa et al. 1968; D'Souza 1970; De Vries 1971, 1972; Welsh and Klatt 1971; Fritz and Lukaszewski 1989). The issue of genetic purity during seed production was less investigated for this species, probably due to the fact that small isolation distances during breeding programs usually avoided unacceptable level of contamination. Nevertheless it was addressed more recently and provides additional information about outcrossing in wheat, which depends on varieties (e.g. Hucl and Matus-Cadiz 2001) and tends to decrease with distance (see Waines and Hegde 2003 for review).

Several authors recognise the need for close-to-reality experimental designs, particularly in respect to the relative sizes of the pollen donor and acceptor plots (Stone 1994; Feil and Schmid 2001) and to the eventuality of unexpected results due to the existence of scale dependent higher order ecological interactions which are intrinsically difficult to test (Wolfenbarger and Phifer 2000). The objective of this study was to evaluate the distances from an agricultural scale wheat field, at which crop pollen was still effective in fertilising *Ae. cylindrica* ovules under competition with auto-pollen and produce viable F1 hybrids. The research is aimed at producing some base line data for decision making, particularly in relation to isolation distances of the wild populations from commercially grown GM wheat.

Materials and Methods

Plant Material: *Ae. cylindrica* seeds were collected during summer 2001 from a natural population in Sierre, canton Valais, Switzerland (Guadagnuolo et al. 2001). Seeds from distinct mother plants were sown in mid-October in individual pots. The young plant rosettes were left over winter for vernalisation at the Botanical Garden of Neuchâtel. As pollen source we used agronomically grown winter wheat fields. The first field (field 1) had a surface of approximately 1.5 hectares and was cultivated with the Swiss wheat variety 'Titlis'. In the second field (field 2), which was privately own by a farmer, approximately one hectare of the French wheat variety 'Galaxie' was grown. Although of a different wheat variety than field 1, this field was chosen because the situation allowed planting *Ae. cylindrica* plots at its edge. Both fields were located at the Swiss Federal Research Station for Agroecology and Agriculture in Reckenholz near Zürich, Switzerland (FAL Reckenholz).

Experimental Design: In March 2002, 262 *Ae. cylindrica* plants were transplanted in plots at 0, 1, 5, 10 and 25 m from wheat field 1 direction South, and in wheat field 2, 213 *Ae. cylindrica* plants were transplanted at 1, 5, 10 and 25 m direction East (Table 1). Zero m plots corresponded to plants within the wheat field (only in field 1, in field 2 it was not possible to plant 0 meter plots due to logistic problems with the owning farmer), whereas all the other plots were planted in two oat fields adjacent to field 1 and 2, direction South and East, respectively. In each location two parallel plot series were planted designated a and b. A distance of 120 m or more was left between the parallel series. The prevailing winds on the Swiss plateau blow from West and North, so that our plots were situated downwind from the pollen source. A plot consisted in two rows, of 13 and 12 *Ae. cylindrica* plants respectively, parallel to the edge of the wheat field (perpendicular to the pollen flow), except for the plots at 25 m, which included 31 or 32 plants distributed on three rows to maximise sensitivity at the biggest distance. In order to minimise *Ae. cylindrica* – *Ae. cylindrica* interactions, a distance of 50 centimetres was left between plants and rows. All seeds produced by the *Ae. cylindrica* plants were collected, and sown in multipots, at the botanical garden. Most *Ae. cylindrica* mother plants produced a reduced amount of tillers and seeds. In plants that produced a very large amount, 30 offspring seeds were sown.

Seeds were sown in autumn 2002 and left in outdoor beds untill maturity. The surviving offspring was screened for morphological hybrids, typically characterised by an intermediate phenotype compared to the parental species (Guadagnuolo et al. 2001). Hybridisation frequencies per plot are expressed as:

% hybrids = (number of hybrid plants per plot / grown offspring plants per plot) x 100

Symmetrical 95% exact binomial confidence intervals were calculated for positive values, while one tail 95% confidence interval was applied to zero values.

Table 1. Experimental design and hybridisation rates at various distances from two distinct wheat fields. Symmetrical 95% exact binomial confidence intervals were calculated for positive values, while one tail 95% confidence interval was applied to zero values. Confidence intervals are indicated in within brackets.

plot	No. plants a	b	seeds sown a	b	survived offspring a	b	No. hybrids a	b	% hybrids a		b	
field 1												
0m	25	25	414	787	316	633	10	2	3.16	(1.53-5.74)	0.32	(0.04-1.14)
1 m	25	25	684	769	346	475	1	0	0.29	(0.1-1.60)	0	(0-0.63)
5 m	25	25	645	821	298	580	0	0	0	(0-1.00)	0	(0-0.52)
10 m	25	25	691	792	359	476	0	0	0	(0-0.83)	0	(0-0.63)
25 m	31	31	1003	980	710	661	0	0	0	(0-0.42)	0	(0-0.45)
field 2												
1 m	25	25	730	773	423	577	0	0	0	(0-0.71)	0	(0-0.52)
5 m	25	25	750	802	423	490	0	0	0	(0-0.71)	0	(0-0.61)
10 m	25	25	764	714	413	435	0	0	0	(0-0.72)	0	(0-0.69)
25 m	31	32	947	979	602	583	0	0	0	(0-0.51)	0	(0-0.51)

field 1: 'Titlis'. field 2: 'Galaxie'. a, b: repetitions within each field

DNA Analyses: Total DNA was extracted from fresh leaves of each of the morphologically distinguished hybrid plants, from a bulk of five individuals of wheat 'Titlis' and 'Galaxie' and from each *Ae. cylindrica* mother plant (leaves stored at -80°C). The extraction followed a simple and cheap SDS-Na-acetate protocol (Savova-Bianchi 1996). Although DNA extraction failed in hybrid number 8 (originating from a 0 m plot) it was included in the calculation of the hybridisation rate. Template DNA was suspended in a TE buffer (pH 8) at a final DNA concentration of 20-40 ng/µl, and stored at -20°C. In order to confirm the hybrid nature of the plants scored with morphology, we performed PCRs using oligonucleotide primers IB10f 5'-CTGCTGGGACCCGATGAATTG-3' and IB10r 5'-TGCTGGGACGAAGCGTTTGAC-3' amplifying a specific 902 bp wheat DNA fragment (Schoenenberger et al. 2005). The 25 µl amplification reaction contained 1x PCR mix, 0.2 mM dNTP, 0.2 µM of each primer, 0.03U/µl Taq polymerase (Quiagen AG, Basel) and 20-40 ng template DNA. Amplifications were carried out in a Biometra T3 thermocycler, with the cycling profile: initial denaturation at 94°C for 10 min then 35 cycles of 93°C, 60s; 58°C, 60s; 72°C, 60s. Final extension was at 72°C, 10 min. PCR products were mixed with 1/5 vol loading buffer and loaded onto 1.5% (w/v) agarose gels stained with ethidium bromide. Electrophoresis was carried out at 100V.

Meteorological Data: All data were obtained from the regional office of MeteoSwiss in Geneva, Switzerland's national weather and climatology service. Data used were collected at 2 m above soil by an automatic observation post, located in Reckenholz, at less than 500 m from both experimental fields. Data refer to the period from May 27th to June 15th 2002, during which pollination occurred. We selected the most relevant variables for pollen flow and pollen viability, namely wind speed and direction, precipitation, temperature and relative humidity (Goss 1968; D'Souza 1970; Khan et al. 1973). For wind speed and main wind direction we retained only data registered during the time gap from 0740 to 1140 h, during which wheat releases most of its pollen (De Vries 1972). Precipitation values are summed up between 0540 and 1140 h.

Results

In 2002, flowering periods of wheat and *Ae. cylindrica* at the field locations were overlapping and *Ae. cylindrica* flowered for a longer time than wheat. In field 1, *Ae. cylindrica* started to flower earlier than 'Titlis', the peak of maximum pollen shed of 'Titlis' was around June 6-7th. In field 2, *Ae. cylindrica* started to flower later than 'Galaxie', which starded heading on May 22 and sheding pollen around the 27th (Table 2). In the two fields *Ae. cylindrica had the same phenology.* 'Galaxie' started to flower about 1 week earlier than 'Titlis'.

Of the 262 transplanted *Ae. cylindrica* mother plants in field 1, 255 survived vegetation period, the dead plants scattered throughout several plots. Seeds produced by the surviving plants were collected and 7586 of them were sown. Survival of the offspring was 64%, the low value is due to the fact that a storm flattened a portion of the oat field in which receptor plots were located during seed maturation, and some *Ae. cylindrica plants had to be collected prematurely, resulting in some* immature seeds, unable to germinate.

Among the 4854 plants which germinated we detected 13 hybrids: 12 from 0 m plots and one from a 1 m plot (Table 1). Hybridisation frequencies per plot, were 3.16% (1.53-5.74 %, for a 95% confidence interval) and 0.32% (0.04-1.14%) in the two 0 m plots and 0.29% (0.1-1.60%) in the first 1 m plot. In all other plots, situated at 1, 5, 10, 25 m from the wheat field edge no hybrid was produced. A wheat-specific DNA marker, absent in *Ae. cylindrica*, was amplified by PCR in the hybrids (Fig. 1). In field 1, by June 7th most wheat plants had exserted stamens with abundant pollen shed. Simultaneously we observed *Ae. cylindrica* plants with widely open spikelets. By June 15th wheat had finished flowering, and by the 18th we could not find any fresh stamen including on the rare plants or spikes that were delayed in flowering. *Ae. cylindrica* kept producing some smaller sized tillers and spikes after June 18th.

In field 2, 204 *Ae. cylindrica* mother plants survived vegetation period, 9 plants died. All seeds were collected. Of the 6459 sown seeds, 3946 produced adult plants (61% survival), and no hybrid was detected among them. 'Galaxie' started flowering about 3 days before *Ae. cylindrica*.

Wind during flowering period blew predominantly from two directions: South-Western to Western (10 days) and Northern to North-Eastern (9 days) sectors (Table 2). Our experimental *Ae. cylindrica* plots were situated South and East of the wheat fields, i.e. downwind. Wind speed in the morning was always above 10 km/h, often above 20 km/h. Rain, which limits pollen flow, fell on 27.05 (4 mm), 28.05 (0.1 mm), 7.06 (0.9 mm) and on 9.06 (2.6 mm). No exceptional meteorological phenomena like storms or extremely high/low temperatures were recorded.

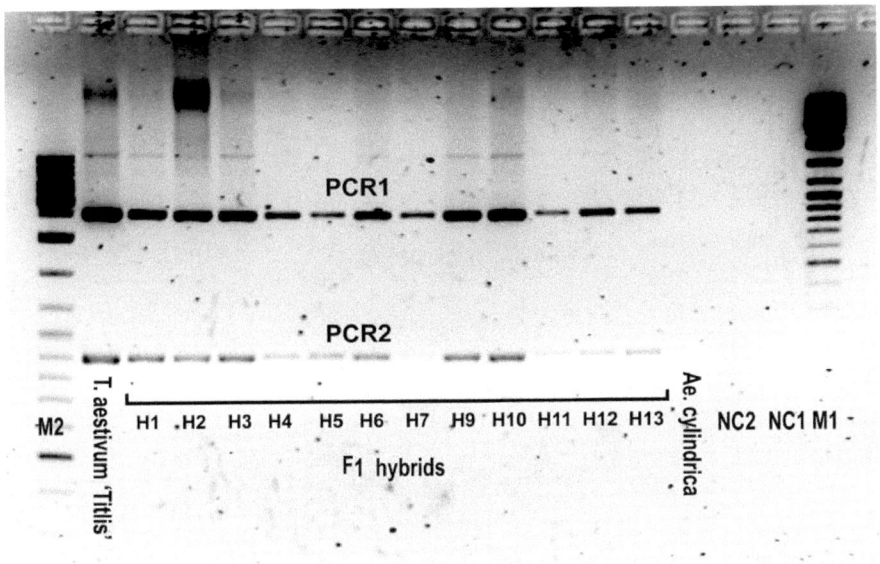

Figure 1. Two repetitions of a PCR loaded at different times onto the same agarose gel showing the same 902 bp DNA marker specific to wheat and present in the hybrids. M:100bp DNA Ladder, NC: negative control (pcr mix without template DNA)

Table 2. Meteorological data during pollination

plants flowering	date 2002	mean temperature °C	precipitations mm	relative humidity ‰ at 11.40 h	wind speed km/h	main wind direction
Gal	27.5	9.1	4	891	>10	SE
Gal	28.5	10.4	0,1	656	>20	SW
Gal	29.5	11.6	0	709	>30	W
Gal/Ae. cyl	30.5	14.3	0	530	>10	N
Gal/Ae. cyl	31.5	16.1	0	436	>10	NE
Gal/Ae. cyl	01.6	17.4	0	465	>20	NE
Gal/Ae. cyl	02.6	18.1	0	480	>20	NE
Gal/Ae. cyl	03.6	15.8	0	587	>10	N
Gal/Ae. cyl	04.6	19.6	0	574	>10	N
Titl/Ae. cyl	05.6	17.2	0	589	>20	W
Titl/Ae. cyl	06.6	15.9	0	574	>20	N
Titl/Ae. cyl	07.6	13.9	0,9	750	>30	SW
Titl/Ae. cyl	08.6	14.8	0	546	>10	NE
Titl/Ae. cyl	09.6	14.7	2,6	798	>10	N
Titl/Ae. cyl	10.6	13	0	571	>40	W
Titl/Ae. cyl	11.6	14.3	0	443	>20	SW
Titl/Ae. cyl	12.6	17.5	0	467	>20	W
Titl/Ae. cyl	13.6	20	0	509	>20	SW
Titl/Ae. cyl	14.6	21.7	0	416	>10	W
Ae. cyl	15.6	23.3	0	405	>10	W

Gal: 'Galaxie'; Titl: 'Titlis', Ae. cyl: Ae. cylindrica

Discussion

Natural *Ae. cylindrica* x *T. aestivum* hybridisation, which is the preferential direction in which crop-to-weed gene flow occurs, has been shown to be possible in Switzerland at a frequency of 1 and 7%, depending on the origin of the *Ae. cylindrica* population (Guadagnuolo *et al.*, 2001). In Oregon (USA) it occurs in wheat fields, at a rate of 0 to 8% depending on the field location (Morrison *et al.*, 2002b). Hybridisation frequencies at 0 m (3.16% and 0.32%) are within the range of previously observed values.

The factors most influencing wheat pollen viability are temperature and relative humidity. High temperatures shorten the longevity as well as low and high relative humidities (Goss,

1968; Welsh & Klatt, 1971). Obermayer (1916) observed that low temperatures had a negative effect on flowering only if combined with humid, misty and windy weather. Optimum conditions for pollen viability are relatively cold temperatures and medium humidities. Under optimal field conditions (20°C, 60% relative humidity) wheat pollen stays viable for about half an hour (D'Souza, 1970). Fritz and Lukaszewski (1989) observed that wheat pollen stays highly viable for about 20 min following anther dehiscence. Viable wheat pollen could have travelled airborne for a much longer distance than the maximum distance of 25 m in our experiment, if we consider a longevity of about 20 min.

'Titlis' in field 1 was the pollinator parent of all detected hybrids. Flowering periods of both species overlapped largely. Conditions for pollination were optimal during flowering in this field. In fact, around the 6-7th June, corresponding to the maximum pollination intensity in field 1, the days were fresh, with predominant North wind, and mean temperatures around 15°C with little variation between minima and maxima. Compatibility between the two species in field 1 was demonstrated by the presence of hybrids in the progeny of *Ae. cylindrica* from the 0 m plots.

Winter wheat disperses considerable amounts of pollen in the air for only one or two days and the day of maximum pollen shed is between day two and day four after anthesis (Khan *et al.*, 1973). Obermayer (1916) notes that maximum flowering in single plants usually takes place on day two and three, sometimes even on day one after anthesis. In our case the maximum peak of pollen release in field 2 could have occurred just before or on the day of anthesis of *Ae. cylindrica*, which is less synchronous than in the cultivated crop. Cold and rainy conditions on May 27-31, with daily temperature minima as low as 5-8°C (data not shown), followed by dry weather, could have negatively influenced pollination in field 2, moreover wind direction was not appropriate after May 29 (Table 2).

Interestingly, in a greenhouse study no hybrids were produced between *Ae. cylindrica* and 'Galaxie' during experimental mixed pollinations (Guadagnuolo *et al.*, 2001). Differences in hybridisation frequencies may be interpreted as varietal effects of the pollen donor and depend on the field location and on meteorological conditions. In fact, substantial differences exist between varieties in the amount of pollen shed (Joppa *et al.*, 1968; De Vries, 1972). However, no data exist for the varieties used in the present study.

Even though distance is by far the most important factor affecting pollen migration (Wagner & Allard, 1991), the question of quantifying hybridisation between *Ae. cylindrica* and bread wheat as a function of distance has been addressed here for the first time. In Western and Northern Europe, *Ae. cylindrica* colonises mainly disturbed habitats such as railway areas, roadsides or vineyards, and to our knowledge has never been reported in or near wheat fields. Recently an *Ae. cylindrica* population has been discovered in Southern Switzerland on a railway station situated in the middle of agricultural land at about 150 m from the closest wheat field (Schoenenberger & Giorgetti-Franscini, 2004), demonstrating that the possibility of contact between the two species exists in Switzerland.

In the present study we show that wheat pollen that may compete with *Ae. cylindrica* pollen to fertilise its ovules does not fly very far from a field. In fact, we detected one single hybridisation event at 1 m from a wheat field's edge, representing a frequency of 0.29% in that plot. No hybridisation was detected at greater distances, neither at 1 m in field 2. When considering a 95% confidence interval, hybridisation rates are always below 0.9% for the distances greater then 5 meters. The presence of cultivated oat between pollen donor fields and the *Ae. cylindrica* plots probably reduced hybridisation distances, although the *Ae. cylindrica* spikes had the same height as the oat canopy. This corresponds to natural situations, where *Ae. cylindrica* grows mixed with other grasses and dicotyledons. Although based on a limited sample size, we can conclude with a certain confidence that minimum isolation distances to effectively avoid hybridisation are small, of the order of a few meters.

Literature about outcrossing in wheat provides indications of the effective viability of its pollen as a function of distance. Intervarietal outcrossing was detected up to 27 m in highly outcrossing cultivars (Hucl & Matus-Cadiz, 2001). Trace level outcrossing events (≤0,01%) were detected up to 80-100 m and exceptionally at 300 m with a rate of 0.005% (Matus-Cadiz et al. 2004). The only information on distance of interspecific gene flow comes from cultivated durum wheat, where gene flow from bread wheat never exceeded 20 m (Matus-Cadiz *et al.*, 2004). Maximum gene flow rates of 0.11-0.19% have been detected at 1 m distance.

Outcrossing in wheat occurs to greater distances than hybridisation. Comparing the data on intra- and interspecific gene flow in wheat, as reviewed by Waines and Hegde (2003) and Hegde and Waines (2004), it appears that, in natural conditions, outcrossing of self-fertilizing cultivars of wheat may be lower than hybridisation rates with wild relatives (Ellstrand, 2003), stressing the necessity to design close-to-reality experiments.

Acknowledgments

We are grateful to Dr. Franz Bigler of the Swiss Federal Research Station for Agroecology and Agriculture in Zürich Reckenholz for allowing the field experiments, to Anouk Béguin for field assistance, to MeteoSwiss, national weather and climatology service, for free meteorological data, and to the team of the Botanical Garden of Neuchâtel for its help in the cultivation of the plants. This project was funded by the National Centre of Competence in Research (NCCR) Plant Survival, a research programme of the Swiss National Science Foundation.

References

Bateman, A. J. 1947. Contamination of seed crops II. Wind pollination. Heredity. 1:235-246.

D'Souza, L. 1970. Untersuchungen über die Eignung des Weizens als Pollenspender bei der Fremdbefruchtung, verglichen mit Roggen, Triticale und Secalotricum. Z. Pflanzenzüchtg. 63:246-269.

De Vries, A. P. 1971. Flowering biology of wheat, particularly in view of hybrid seed production – a review. Euphytica. 20:152-170.

De Vries, A. P. 1972. Some aspects of cross-pollination in wheat (*Triticum aestivum* L) .1. Pollen concentration in the field as influenced by variety, diurnal pattern, weather conditions and level as compared to height of pollen donor. Euphytica. 21:185-203.

Donald, W. W. and A. G. Ogg. 1991. Biology and control of jointed goatgrass (*Aegilops cylindrica*), a review. Weed Technol. 5:3-17.

Ellstrand, N. C. 2003. Dangerous liaisons? When cultivated plants mate with their wild relatives. Baltimore: The Johns Hopkins University Press. Pp.171-178 & 194.

Ellstrand, N. C. and C. A. Hoffman. 1990. Hybridization as an avenue of escape for engineered genes. BioScience. 40:438-442.

Ellstrand, N. C., H. C. Prentice, and J. F. Hancock. 1999. Gene flow and introgression from domesticated plants into their wild relatives. Annu. Rev. Ecol. Syst. 30:539-563.

Feil, B. and J. E. Schmid. 2001. Pollenflug bei Mais, Weizen und Roggen. Aachen: Shaker. 69 p.

Fritz, S. E. and A. J. Lukaszewski. 1989. Pollen longevity in wheat, rye and triticale. Plant Breed. 102:31-34.

Goss, J. A. 1968. Development, physiology, and biochemistry of corn and wheat pollen. Bot. Rev. 34:333-358.

Guadagnuolo, R., D. Savova-Bianchi, and F. Felber. 2001. Gene flow from wheat (*Triticum aestivum* L.) to jointed goatgrass (*Aegilops cylindrica* Host.), as revealed by RAPD and microsatellite markers. Theor. Appl. Genet. 103:1-8.

Hegde, S. G. and J. G. Waines. 2004. Hybridization and introgression between bread wheat and wild and weedy relatives in North America. Crop Sci. 44:1145-1155.

Hucl, P. 1996. Out-crossing rates for 10 Canadian spring wheat cultivars. Can J Plant Sci. 76:423-427.

Hucl, P. and M. Matus-Cadiz. 2001. Isolation distances for minimizing out-crossing in spring wheat. Crop Sci. 41:1348-1351.

Janakiraman, V., M. Steinau, S. B. McCoy, and H. N. Trick. 2002. Recent advances in wheat transformation. In Vitro Cell. Dev. Biol.-Plant. 38:404-414.

Joppa, L., F. McNeal, and M. Berg. 1968. Pollen production and pollen shedding of hard red spring (*Triticum aestivum* L. em Thell.) and durum (*T. durum* Desf.) Wheats. Crop Sci. 8:487-490.

Khan, M. N., E. G. Heyne, and A. L. Arp. 1973. Pollen distribution and seedset on *Triticum aestivum* L. Crop Sci. 13:223-226.

Martin, T. J. 1990. Outcrossing in Twelve Hard Red Winter Wheat Cultivars. Crop Sci. 30:59-62.

Matus-Cadiz, M. A., P. Hucl, M. J. Horak, and L. K. Blomquist. 2004. Gene flow in wheat at the field scale. Crop Sci. 44:718-727.

Morrison, L. A., L. Crémieux, and C. A. Mallory-Smith. 2002a. Infestations of jointed goatgrass (*Aegilops cylindrica*) and its hybrids in Oregon wheat fields. Weed Sci. 50:737-747.

Morrison, L. A., O. Riera-Lizarazu, L. Crémieux, and C. A. Mallory-Smith. 2002b. Jointed goatgrass (*Aegilops cylindrica* Host) x wheat (*Triticum aestivum* L.) hybrids: hybridization dynamics in Oregon wheat fields. Crop Sci. 42:1863-1872.

Moser, D., A. Gygax, B. Bäumler, N. Wyler, and R. Palese. 2002. Liste Rouge des fougères et plantes à fleur menacées de Suisse. Bern: Office Fédéral de l'Environnement des Forêts et du Paysage; Chambésy: Centre du Réseau Suisse de Florstique; Conservatoire et Jardin Botaniques de la Ville de Genève. 41, 120 p.

Obermayer, E. 1916. Untersuchungen über das Blühen und die Befruchtung von Winterroggen und Winterweizen. Z. Pflanzenzüchtg. 4:347-403.

Rajhathy, T. 1960. Continuous spontaneous crosses between *Aegilops cylindrica* and *Triticum aestivum*. Wheat Inf Ser. 11:20.

Savova-Bianchi, D. 1996. Evaluation of gene flow between crops and related weeds: risk assessment for releasing transgenic barley (*Hordeum vulgare* L.) and Alfalfa (*Medicago sativa* L.) in Switzerland. PhD thesis, University of Neuchâtel, Switzerland.

Schoenenberger, N., F. Felber, D. Savova-Bianchi, and R Guadagnuolo. 2005. Introgression of wheat DNA markers from A, B and D genomes in early generation progeny of *Aegilops cylindrica* Host x *Triticum aestivum* L. hybrids. Theor Appl Genet. 111:1338-1346.

Schoenenberger, N. and P. Giorgetti-Franscini. 2004. Note floristiche ticinesi: la flora della rete ferroviaria con particolare attenzione alle specie avventizie. Parte II. Boll. Soc. Tic. Sci. Nat. 92:97-108.

Slageren, M. W. van. 1994. Wild wheats: a monograph of *Aegilops* L. and *Amblyopyrum* (Jaub. & Spach) Eig (Poaceae). Wageningen: Wageningen Agricultural University Press; Aleppo: ICARDA. 204-208, 512 p.

Stone, R. 1994. Large plots are next test for transgenic crop safety. Science. 266:1472-1473.

Stone, A. E. and T. F. Peeper. 2004. Characterizing jointed goatgrass (*Aegilops cylindrica*) x winter wheat hybrids in Oklahoma. Weed Science. 52:742-745.

Wagner, D. B. and R. W. Allard. 1991. Pollen migration in predominantly self-fertilizing plants: barley. J. Hered. 82:302-304.

Waines, J. G. and S. G. Hegde. 2003. Intraspecific gene flow in bread wheat as affected by reproductive biology and pollination ecology of wheat flowers. Crop Sci. 43:451-463.

Watanabe, N. and T. Kawahara. 1999. *Aegilops* species collected in California and Oregon, USA. Wheat. Info. Serv. 89:33-36.

Welsh, J. R. and A. R. Klatt. 1971. Effects of temperature and photoperiod on spring wheat pollen viability. Crop Sci. 11:864-865.

Wolfenbarger, L. L. and P. R. Phifer. 2000. Biotechnology and ecology - The ecological risks and benefits of genetically engineered plants. Science. 290:2088-2093.

Zemetra, R. S., J. Hansen, and C. A. Mallory-Smith. 1998. Potential for gene transfer between wheat (*Triticum aestivum*) and jointed goatgrass (*Aegilops cylindrica*). Weed Sci. 46:313-317.

CHAPTER 2

Introgression of wheat DNA markers from A, B and D genomes in early generation progeny of *Aegilops cylindrica* Host x *Triticum aestivum* L. hybrids[2]

---

[2] Published in Theoetical and Applied Genetics (2005) 111: 1338-1346
DOI 10.1007/s00122-005-0063-7

Abstract

Introgression from allohexaploid wheat (*Triticum aestivum* L., AABBDD) to allotetraploid jointed goatgrass (*Aegilops cylindrica* Host, CCDD) can take place in areas where the two species grow in sympatry and hybridise. Wheat and *Ae. cylindrica* share the D genome, issued from the common diploid ancestor *Aegilops tauschii Coss*. It has been proposed that the A and B genome of bread wheat are secure places to insert transgenes to avoid their introgression into *Ae. cylindrica* because during meiosis in pentaploid hybrids, A and B genome chromosomes form univalents and tend to be eliminated whereas recombination takes place only in D genome chromosomes.

Wheat random amlpified polymorphic DNA (RAPD) fragments, detected in intergeneric hybrids and introgressed to the first backcross generation with *Ae. cylindrica* as the recurrent parent and having a euploid *Ae. cylindrica* chromosome number or one supernumerary chromosome, were assigned to wheat chromosomes using Chinese Spring nulli-tetrasomic wheat lines. Introgressed fragments were not limited to the D genome of wheat, but specific fragments of A and B genomes were also present in the BC1. Their presence indicates that DNA from any of the wheat genomes can introgress into *Ae. cylindrica*. Successfully located RAPD fragments were then converted into highly specific and easy-to-use sequence characterised amplified regions (SCARs) through sequencing and primer design. Subsequently these markers were used to characterise introgression of wheat DNA into a BC1S1 family. Implications for risk assessment of genetically modified wheat are discussed.

Introduction

Gene flow by sexual reproduction from crops to their wild relatives, represents one of the main concerns about commercial cultivation of genetically modified plants. It occurs through the production of first generation hybrids, followed by several backcrosses. The consequences of gene flow from conventional or transgenic crops may be severe as increased weediness and loss of biodiversity of wild taxa (Ellstrand 2003).

The wild species jointed goatgrass (*Aegilops cylindrica* Host, Poaceae) is native to a wide area from central Asia to the Mediterranean (Van Slageren 1994). It has 2n = 4x = 28 chromosomes, an allotetraploid origin, and a genomic formula notation of CCDD. *Ae. cylindrica* hybridises naturally with allohexaploid wheat (*Triticum aestivum* L.) (Van Slageren 1994), which has 2n = 6x = 42 chromosomes, and a AABBDD genome. The reason why intergeneric crosses between the two species are possible has been attributed to the fact that they share the D genome ancestor (Zemetra et al. 1998), although at least

12 *Aegilops* species acting as female parents form natural hybrids with hexaploid wheat and not all of them have a common genome with it (Van Slageren 1994). *Ae. cylindrica* x *T. aestivum* hybrids are male sterile but produce viable BC1 seeds at low frequency through backcrossing to either of the parental species (Rajhathy 1960; Guadagnuolo et al. 2001; Morrison et al. 2002b).

Ae. cylindrica was introduced as an adventive to the USA in the late 19th century and became a serious agricultural weed in North America, particularly in winter wheat fields (Donald and Ogg 1991). In Oregon (USA) wheat fields, *Ae. cylindrica* was found to be the predominant female parent of F1 hybrids with wheat (Morrison et al. 2002b). In fact, in a population of *Ae. cylindrica* growing within or on the edge of a wheat field and exposed to a massive wheat pollen load, wheat was the predominant male parent of the BC1 generation. The situation is different in Europe and in areas where crop rotations are practiced: the hybrids would grow within the *Ae. cylindrica* population with little pollen pression from the crop. The BC1 would thus most probably be produced with *Ae. cylindrica* acting as pollen parent, as was experimentally shown by Guadagnuolo et al. (2001). Experimental research in the USA focused exclusively on *T. aestivum* x *Ae. cylindrica* hybrids and their backcross derivatives (reviewed in Hegde and Waines 2004). In these hybrids, D genome chromosomes can pair successfully during meiosis, allowing gene recombination, while wheat A and B genome chromosomes remain univalents and tend to be eliminated as the number of backcross generations with *Ae. cylindrica* increases (Zemetra et al. 1998). Consequently, the risk of transgene spread from wheat to *Ae. cylindrica* could be reduced by placing the transgene on the A or B genome of wheat. Analysis of 14 wheat microsatellites, one for each D genome chromosome arm, in *T. aestivum* x *Ae. cylindrica* hybrids and BC1 and BC2 with *Ae. cylindrica* revealed that most of them were inherited in a Mendelian fashion (Kroiss et al. 2004). Two markers in BC1 (located on chromosome arms 5DS and 7DS), and three in BC2 (on 1DS, 2DL, 7DS), were retained at a higher proportion than expected. In fact, D genome chromosomes tend to remain 14 in number in every generation after hybridisation, whereas C genome chromosomes will increase to 14 as the backcrossing with *Ae. cylindrica* continues (Wang et al. 2000). The stable chromosome number and genome composition will then be 14 D chromosomes and 14 C chromosomes, as in *Ae. cylindrica*. In BC2S2 (BC2 selfed twice) plants having 29, 30, or 31 chromosomes, it was shown by genomic in situ hybridisation (GISH) that the extra chromosomes were retained from the A and B genomes of wheat, but these chromosomes were not present in homologous pairs and thus not expected to be stable over generations (Wang et al. 2000). Nevertheless, one translocation of a chromosome part originating from the A or B genome was found in a BC2S2 plant. The translocated segment is likely to be maintained in the *Ae. cylindrica* background (Wang et al. 2000). On the other hand, it was shown that genetically engineered herbicide resistance using the *bar* gene could be transferred from the wheat cultivar Bobwhite to manually produced *Ae. cylindrica* backcross populations if it was located on the D genome, whereas no resistant plants were found in a reduced sample of BC1 individuals having wheat with the *bar* gene on the B genome as the pollen parent in the F1 cross (Lin 2001).

In a previously reported study, *Ae. cylindrica* x *T. aestivum* hybrids and BC1 with the wild species were produced in field experiments. Several wheat-specific RAPD fragments were inherited up to the BC1 generation, with the analysed plants having euploid *Ae. cylindrica* chromosome number of 28 or with one additional chromosome (Guadagnuolo et al. 2001). The aim of the present study was to assign the introgressed RAPD fragments to wheat chromosomes and to verify whether they would be maintained in the *Ae. cylindrica* background following self-fertilisation and backcrossing with the wild species. Furthermore, the RAPD fragments were sequenced and converted into highly specific and easy-to-use sequence characterised amplified region markers (SCARs, Paran and Michelmore 1993). These SCAR markers will be useful to trace introgressed wheat DNA along the generations.

Materials and Methods

Plant material: *Aegilops cylindrica* Host x *Triticum aestivum* L. hybrids and first backcross generations (BC1) with *Ae. cylindrica* as the pollen parent were produced under open pollination conditions in a field experiment (Guadagnuolo et al. 2001). Additionally, a BC2 family, i.e. the descendants of one of the BC1 individuals mentioned above, grown under open pollination conditions together with *Ae. cylindrica*, and their probably selfed offspring (BC2S1) were analysed. As the BC2 individuals were growing isolated from each other and from *Ae. cylindrica,* we consider their offspring as the product of self-fertilisation. In a second step, an introgressive series up to a BC1S1 family was generated. This latter was produced from a manual cross between *Ae. cylindrica* from a natural population in Sierre, canton Valais, Switzerland (Guadagnuolo et al. 2001), and the wheat variety Bobwhite line 5 (Abranches et al. 2000) as pollen donor, backcrossed with *Ae. cylindrica* under open pollination conditions. Bobwhite 5 was obtained by Eva Stoger of the Institute for Biology of the RWTH Aachen, Germany.

For chromosome assignment of RAPD and SCAR fragments, we used twenty-one nullisomic-tetrasomic (NT) wheat lines of the variety Chinese Spring (CS) originally produced by E. R. Sears (Sears 1966). DNA extracts of the NT lines and euploid CS were obtained from Beat Keller of the Institute of Plant Biology of the University of Zürich, Switzerland. Moreover, some wheat relatives and ancestor species having a fraction of the wheat hexaploid genome were analysed to investigate whether introgressed RAPD markers could be found in the genome donor species too. The investigated species were *Triticum monococcum* L., *Triticum urartu* Thüm. (genome formula AA), *Triticum turgidum* L., *Triticum dicoccum* Schrank (AABB), and *Aegilops tauschii* Coss. (DD).

Chromosomes were counted in root meristemes cells as described in Savova et al. (1996).

RAPD analysis and chromosome assignment: Genomic DNA was extracted from fresh or frozen (-80°C) leaf tissue using the DNeasy Plant Mini Kit (Qiagen, Basel, Switzerland), following the manufacturer's protocol. Final DNA concentration was adjusted to 20-40 ng/µl, and stored at -20°C. RAPD analysis was carried out as described in Guadagnuolo et al. (2001), using primers OPB6, 8, 10, and OPP6, 8, 9, 14 (Operon Technologies, Alameda, CA, USA). Sixteen wheat-specific RAPD fragments were detected in the F1 hybrids; nine of them were present in the BC1 generation (Guadagnuolo et al. 2001). The 16 RAPD fragments were amplified in CS NT lines for chromosome assignment. A BC2 family and their BC2S1 offspring were analysed by RAPD analysis in order to test introgression to further generations.

Direct sequencing of purified RAPD products: Seven of the RAPD fragments successfully localised on wheat chromosomes were chosen for sequencing (Table 1: DP9, GP8, GB10, IB10, FP8, YP8, D1P9). The choice was based on intensity and distinctiveness of the RAPD band, on its specificity to wheat, and on its persistence in *Ae. cylindrica x T. aestivum* backcross derivatives. The fragments were purified from the agarose electrophoresis gel using the QIAquick Gel Extraction Kit (Qiagen), following the manufacturer's protocol. Purified RAPD fragments were directly sequenced as described by Hernandez et al. (1999). However, instead of using PCR products flanked by two different oligonucleotides as in that report, we used a purified fragment generated by only one 10-mer RAPD primer and that primer for a first cycle sequencing reaction. Usually one strand of the sequence was preferentially amplified, allowing the design and synthesis of a strand specific ≥20mer oligonucleotide (Microsynth, Balgach, Switzerland). This primer was then used in another round of cycle sequencing with the purified RAPD fragment to obtain a clean sequence of it. Reverse primers were designed to sequence the complementary strand.

Cycle sequencing was performed using the dideoxy chain dye-termination method with an ABI PRISM Big Dye Terminator Cycle Sequencing Ready Reaction Kit (Applied Biosystems, Foster City, CA, USA) in a Biometra T3 thermocycler. Cycling parameters were 25 cycles of 20 sec at 96°C, 20 sec at 50-60°C annealing temperature depending on the primer used, and 4 min at 60°C. Cycle sequencing was performed in 5 or 10 µl reaction volume, depending on the success of cycling reactions. The products were then purified using the ethanol/EDTA/Sodium acetate precipitation method suggested by the sequencing kit's manufacturer. The purified sequencing products were resuspended in 12 µl TSR (Applied Biosystems), and run in an ABI 310 Automated Sequencer (Applied Biosystems). Basecalling was checked manually against the electropherograms, and edited if necessary, using the Sequence Navigator software (Applied Biosystems). Checked and eventually edited sequences were aligned using the Sequence Alignement Editor software (Andrew Rambaut, University of Oxford), in order to obtain the entire sequence of the RAPD fragments.

In one case (fragment DP9), both strands were equally amplified resulting in superposed electropherograms, even when the annealing temperature was increased up to 60°C. The corresponding RAPD primer elongated by each A,T,G and C nucleoside was synthesised (Microsynth) and PCR conditions optimised using each possible primer pair combination in a Biometra Tgradient thermocycler with a gradient of annealing temperatures of 45-59°C. PCR products were checked on ethidium bromide stained 1,4% (w/v) agarose gels. The optimised PCR conditions were used to perform cycle sequencing. Sequences have been deposited in GenBank; accession numbers are indicated in Table 1.

Table 1. Chromosome assignment of specific wheat fragments detected in hybrids and backcrosses with Ae. cylindrica and marker conversion

RAPD Fragment	Chrom. location RAPD	Presence in hybrids	Presence in BC_1	Sequencing (sequence size)	GenBank accession No.	Wheat-specific SCAR	Chrom. location SCAR
DP9	6A	Yes	Yes	Yes (644 bp)	AY842011	Yes	6A
DB6	1B	Yes	Yes	No	-	-	-
GP8	2B	Yes	Yes	Yes (950 bp)	AY842009	Yes	No/7D
HB8	5B	Yes	No	No	-	-	-
GB10	5B	Yes	Yes	Yes (673 bp)	AY842010	Yes	5B
IB10	7B	Yes	Yes	Yes (903 bp)	AY842008	Yes	No
FP8	7B	Yes	No	No	-	-	-
YP8	6D	Yes	Yes	Yes (402 bp)	-	No	-
D1P9	6D	Yes	Yes	Yes (764 bp)	AY842012	Yes	6D

Chrom.: chromosome

SCAR primer design and PCR: Sequences of the investigated RAPD fragments were used to design specific SCAR primer pairs 20-21 bases long. The primers were located internally or were superposed to the RAPD primer binding site resulting in PCR products usually slightly shorter than the RAPD fragment. Specific PCR was performed in 25 µl reactions containing 1x PCR mix, 0.2 mM dNTP, 0.2 µM of each primer, 0.03U/µl Taq polymerase (Qiagen) and 20-40 ng template DNA. Amplifications were carried out in a Biometra T3 thermocycler, with the cycling profile: initial denaturation at 94°C for 10 min, then 35 cycles of 93°C for 60 sec; 55-59°C for 60 sec, depending on the primer pair (Table 2); 72°C for 60 sec. Final extension was at 72°C for 10 min. PCR products were mixed with 1/5 vol loading buffer and loaded onto a 1.5% (w/v) agarose gel, stained with ethidium bromide. Electrophoresis was carried out at 100V. Recovered SCAR markers were tested for specificity to wheat, for maintenance of introgressive pattern and for chromosome location. Afterwards, we analysed an introgressive series up to BC1S1 using the SCAR markers.

Results

Chromosome Assignment: Among the 16 specific wheat RAPD fragments inherited by *Ae. cylindrica* x *T. aestivum* hybrids, nine were assigned to wheat chromosomes, and the remaining seven were amplified in all NT lines. It was therefore impossible to determine their chromosome location. Among the nine wheat fragments detected in BC1 plants (Guadagnuolo et al. 2001), seven were assigned to wheat chromosomes of the A, B and D genomes (Table 1). Four of these RAPD fragments originate from the B genome of wheat, two from the D genome and one from the A genome (Figure 1). None of these fragments were amplified in the ancestor species and other wild relatives possessing a part of the wheat genome, indicating that each of the RAPD fragment arose in allopolyploid wheat, either because of mutations at one or both of the RAPD primer binding sites, because of sequence elimination in the polyploid, or by chromosome rearrangements.

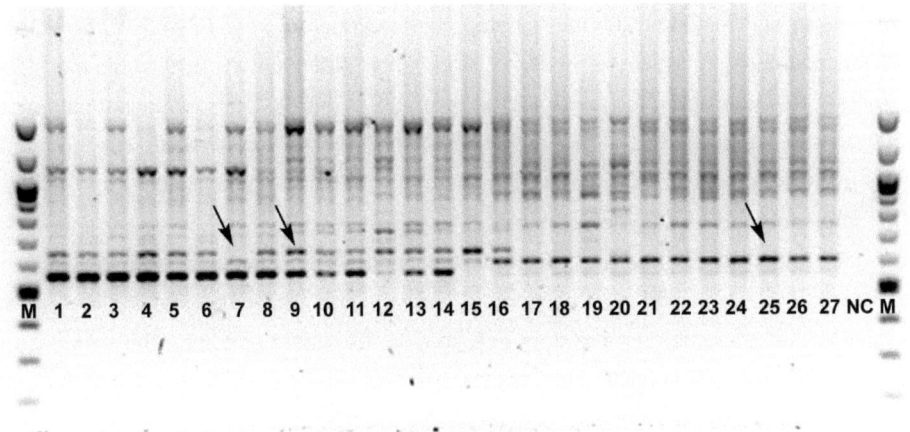

Figure 1: RAPD amplification with primer OPP9 in euploid *T. aestivum* var. Chinese Spring (lane 1) and its nulli-tetrasomic lines for the seven A genome chromosomes (lanes 2-8), *T. aestivum* var. Arina (lane 9), *Ae. cylindrica* (lane 25), *Ae. cylindrica* x *T. aestivum* hybrids (lanes 10-14), BC_1 (lanes 15-21), BC_2 (lanes 22-24), and BC_2S_1 plants (lanes 26-27). Arrows indicate fragment DP9 localised on chromosome 6A, present in wheat and absent in *Ae. cylindrica*. M: 100bp DNA Ladder, NC: negative control

Development of SCARs: Direct sequencing of the RAPD fragment using the original single 10mer primer was successful in 5 of the 7 fragments chosen. These first sequences usually had a quite high background noise represented by the sequence of the complementary strand, because in the RAPD technique, both primer binding sites consist of the same sequence. However, cycle sequencing reactions using these fragments produced directly a readable sequence allowing the design of more specific 20mer primers. These strand-specific primers were then used to sequence the whole fragment. One fragment (DP9), always produced a double sequence. It could only be sequenced using the corresponding RAPD primer elongated by one nucleoside (Cytosine), which made the primer strand-specific, thus allowing it to produce sequence in only one direction. Fragment FP8 always produced a double sequence too. The procedure of RAPD primer elongation by each nucleotide was not carried out because this fragment was not present in BC1 plants. This fragment was thus discarded from further investigations (Table 1). Based on the obtained sequences, six 20-21mer primer pairs (Table 2) were synthesised and tested on the investigated plant material in order to verify if the SCARs maintained the same pattern as the RAPD fragments. Unfortunately, by the time the SCAR primers were developed, the original DNA extracts of the BC1 from Guadagnuolo et al. (2001) were degraded and did not amplify any more. Five of the newly developed primer pairs maintained specificity to wheat. One fragment (YP8), also amplified in *Ae. cylindrica*. The sequences of the fragments amplified by the SCAR primer set in wheat and in *Ae. cylindrica* were almost identical, with some polymorphic nucleotides (data not shown). Moreover, the sequence was polymorphic even within a single wheat DNA extract indicating that it existed in more than one copy in the wheat genome. The sequence of fragment YP8 was not submitted to GeneBank because of its polymorphic nature in wheat. Marker YP8 was discarded from further analysis because of the loss of specificity to wheat. The remaining five newly developed SCARs were then tested on CS NT lines to verify their chromosome location. Chromosome location was successfully identified and identical to the corresponding RAPD fragments for three of these markers, one for each wheat A, B and D genome. For two fragments, chromosome assignment was impossible after marker conversion. Fragment IB10 was amplified in all 21 NT lines, suggesting that the newly designed primers amplified at least two different regions on different chromosomes of the wheat genome. The GP8 fragment contains a 107 bp tandem repeat, the second repeat having two mutated nucleotides and one deleted compared to the first. The primer pair designed external to the repeated zone of this sequence amplified two fragments, one of the expected 950 bp length (same as the corresponding RAPD fragment) and an additional fragment of about 800 bp. The additional fragment was located on the wheat chromosome 7D, whereas the 950 bp fragment could not be localised to chromosome. We tested other primer pairs located external to the repeat but internal to the GP8 sequence. They amplified one single band in all NT lines; thus, the chromosome location of GP8 remains unknown.

Database comparisons of the obtained sequences in GenBank's nr databases, using the Basic Local Alignment Search Tool (BLAST), revealed a high degree of sequence homology for DP9, IB10 and YP8 fragments to genes from other organisms such as various diploid *Triticum* and *Aegilops* species, *Triticum turgidum*, *Triticum aestivum*, and other members of the Triticeae tribe like *Hordeum* and *Agropyron*. Most of the identified genes were related to metabolism, storage, structure and even putative disease resistance. Moreover, some of the identified sequences were retrotransposons.

Table 2. SCAR primers, size of the amplified fragment, and annealing temperatures used for specific PCR

name	sequence (5'→3')	position in RAPD fragment (5'→3')	size SCAR (bp)	annealing temperature (°C)
DP9F	ACGGCAATTCTTTATGGAAGT	21-41	610	55
DP9R	TAACATTCGATGATGACCGG	611-630		
GP8F	AGATTCAGTTGCACCATCAC	16-35	931	55
GP8R	CGCCCAAGGATAGCAGTCCTT	926-946		
GB10F	TGGGACCAGGATTGTGAGTAC	5-25	662	59
GB10R	GACGTGCACAAGTGGGAGGAA	646-666		
IB10F	CTGCTGGGACCCGATGAATTG	1-21	902	58
IB10R	TGCTGGGACGAAGCGTTTGAC	882-902		
YP8F	ACATCGCCCACTCTCAGAGG	1-20	402	55
YP8R	ACATCGCCCATCTTGATAACG	382-402		
D1P9F	TGGCAACAGGGTAATGATCCC	22-42	737	58
D1P9R	CGCAAAATTTGGTTTAGGGCT	738-758		

DNA marker Introgression: Although several RAPD fragments assigned to wheat chromosomes were detected in the field-produced BC1 plants, none of these fragments were present in the analysed BC2 family and in their BC2S1 offspring (Figure 1). This is not surprising considering the fact that they were all descendants of one single BC1 individual, which unfortunately was no longer available for analysis.

An F1 individual issued from a manual cross, its BC1 with *Ae. cylindrica*, and the selfed BC1S1 offspring were analysed using three SCARs, one for each wheat genome. Markers D1P9 (6D) and DP9 (6A) were introgressed to the BC1S1 generation whereas GB10 (5B) was only present in the hybrid and was not present in the BC1 plant (Table 3). The BC1S1 plants had 28 to 31 chromosomes, i.e. the euploid *Ae. cylindrica* chromosome number and up to 3 additional chromosomes. The BC1S1 individual 5.13.11 had 28 chromosomes and a marker from the wheat chromosome 6A (Figure 2, lane 12).

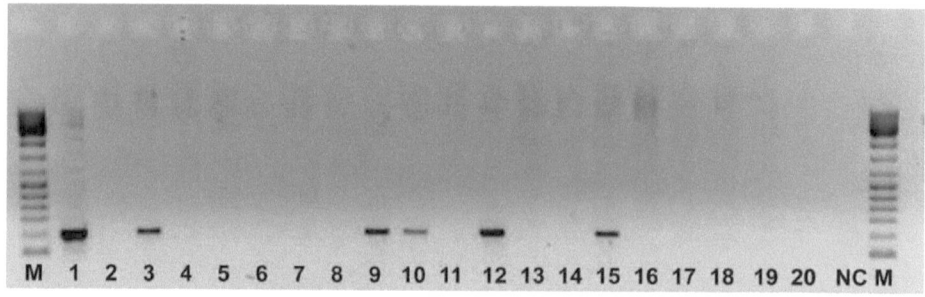

Figure 2: Amplification of the SCAR DP9 in the wheat variety Bobwhite (lane 1), *Ae.cylindrica* x *T. aestivum* BC1S1 (lane 2-19), *Ae.cylindrica* (lane 20). NC: negative control (lane 21), M: 100bp DNA Ladder

Discussion

Marker conversion: A direct sequencing method of RAPD fragments avoids the costly and time-consuming cloning step generally used for SCAR development from RAPD markers (Paran and Michelmore 1993). Cycle sequencing of a RAPD fragment generated using one single primer would be expected always to produce a double sequence since both primer sites are identical; thus, it has been thought that only fragments flanked by two different RAPD primers can be sequenced directly (Hernandez et al. 1999). Here we report a method of direct sequencing with only one RAPD primer which produced directly a readable sequence in five out of seven fragments of interest. For one of the other fragments the RAPD primer was elongated by one nucleotide and a readable sequence was obtained, while for the other fragment no attempt at primer elongation was undertaken. In a case where the flanking nucleotide at the 3' end of the two primer binding sites is the same, one could elongate the primer by two nucleotides; however, the possibility of different primer design grows exponentially with every further nucleotide added to the RAPD primer, and so does the cost of the operation. To explain the high proportion of single-primer RAPD fragments generated in this study that were directly sequenceable, one must assume that the binding of the RAPD primer at each end of the successfully sequenced fragment, or the activity of the polymerase, must have been somewhat different. If template DNA preparation and cycle sequencing reactions are optimised to permit direct sequencing using a single primer RAPD product, the procedure described here is even more time and cost effective as it avoids the step of screening for markers amplified only by two different primers described Hernandez et al. (1999).

The identification of primers and reaction conditions sufficiently specific to produce an amplified product from a marker that is a single fragment, presence/absence polymorphism can enable the amplification product to be directly stained in the PCR tubes, thus avoiding the gel electrophoresis step (Hernandez et al. 2001). This was the case of three out of four successfully located SCARs developed here. The problem of reamplification of untargeted bands (Hernandez et al. 1999) was overcome here as only RAPD fragments already localised on CS NT lines were used for sequencing. For the purpose of detecting introgressed wheat fragments localised on single chromosomes, the outcome of the overall described procedure was rather low; out of nine successfully located RAPD fragments only four useful markers were developed. This is probably due to the complexity of the allohexaploid wheat genome with several triplicated loci on homeologous chromosomes. Paradoxically, increasing PCR specificity by the use of longer oligonucleotide primers decreased genome specificity in three out of six cases. In one case (fragment YP8), PCR with 20mer primers even lost specificity to wheat. Once the primer was elongated homology of the primer at its 3' ends was increased, causing loss of polymorphism. Similarly, a RAPD fragment specific to *Hordeum chilense*, after it was converted to a SCAR, lost its specificity to *H. chilense* and showed amplification in wheat too; mismatch of the RAPD primer rather than sequence polymorphism was considered to be the cause of the RAPD polymorphism (Hernandez et al. 1999). Five of the six successfully sequenced fragments can be used to demonstrate hybridisation between wheat and *Ae. cylindrica*, even though chromosome location of one of these markers (IB10) remains unknown.

An advantage of using RAPD markers for introgression studies is that these markers occur in randomly amplified genome regions, and if sequenced and compared to nucleotide databases they may provide some information on eventual genes or traits introgressed from one species to another.

Table 3. chromosome numbers and presence (+) of specific wheat SCAR markers (chromosome location annotated) in the first three generations of an introgressive series from wheat var. Bobwhite to *Ae. cylindrica*

plant	Chromosome number	D1P9 (6D)	DP9 (6A)	GB10 (5B)
F1				
5.13	35	+	+	+
BC1				
5.13	30	+	+	-
BC1S1				
5.13.1	28	-	-	n.a.
5.13.2	29 + t	+	+	
5.13.3	29	+	-	
5.13.4	29	-	-	
5.13.5	28	+	-	
5.13.6	n.a.	-	-	
5.13.7	30	+	-	
5.13.8	29	+	+	
5.13.9	31	+	+	
5.13.10	28	-	-	
5.13.11	28	+	+	
5.13.12	28	+	-	
5.13.13	30	+	-	
5.13.14	29	+	+	
5.13.15	28	+	-	
5.13.16	28	+	-	
5.13.17	28	-	-	
5.13.18	30	+	-	

n.a.: not analysed. t: telocentric

Marker introgression and implications for transgene flow: Only 9 out of 16 wheat-specific RAPD fragments investigated could be assigned to wheat chromosomes, while the remaining 7 markers were amplified in all 21 CS NT lines. This result is consistent with observation of Devos and Gale (1992) where only a fraction of RAPD fragments could be localised by aneuploid analysis. The unlocalised markers were probably present on more than one wheat chromosome, possibly members of a homeologous group, or the markers consisted of repetitive DNA. Wheat-specific markers detected in partially fertile BC1 plants having 28 or 29 chromosomes (Guadagnuolo et al. 2001) were assigned to the A, B and D genomes of wheat. Only two out of seven wheat-specific RAPD markers were located on the D genome (YP8 and D1P9), whereas four originate from the B genome. In fact, the D genome of wheat is known to be less polymorphic than the A and B genome (Nelson et al. 1995; Cadalen et al. 1997).

There are three mechanisms by which wheat genetic material can be introgressed into *Ae. cylindrica*: (1) recombination of the homologous D genome chromosomes, (2) intergenomic translocations or chromosome rearrangements, and (3) disomic chromosome retention. Recombination of homologous chromosomes is the most frequent mechanism and even though the D genome of wheat and *Ae. cylindrica* originated from different biotypes of *Ae. tauschii* (Badaeva et al. 2002) they pair successfully at meiosis forming 14 bivalents (Zemetra et al. 1998). In fact, inheritance of most microsatellites specific to the D genome of wheat fit the expected Mendelian ratio and were neither preferentially inherited nor lost in *T. aestivum* x *Ae. cylindrica* hybrids, BC1 and BC2 families (Kroiss et al. 2004). Using GISH a translocation from the A or B genome of wheat to an *Ae. cylindrica* chromosome was detected in a BC2S2 plant (Wang et al. 2000). Furthermore, the authors established that in plants with 30 chromosomes the two extra chromosomes were not a homologous pair. If they had been, one would expect the extra chromosome pair to be inherited in a stable manner (Wang et al. 2000). The field-produced BC1 (Guadagnuolo et al. 2001) plants investigated in the present study had 28 or 29 chromosomes; thus, the markers from the A and B genome detected in these plants were on a supernumerary chromosome, on a substitution A or B chromosome, or translocated. The introgressive series issued from a manual *Ae. cylindrica* x *T. aestivum* var. Bobwhite cross contains BC1S1 individuals having 28-31 chromosomes (Table 3). The supernumerary chromosomes may be present in homologous pairs, as this generation was produced by selfing of a BC1 individual having 30 chromosomes. Individual 5.13.9 had 31 chromosomes, meaning that either there was one A or B genome chromosome present in a homologous pair, and thus expected to be stably inherited, or there were three different A or B genome chromosomes present as monosomes, which presupposes the presence of a substituted chromosome in the BC1 plant. Individual 5.13.11 had a euploid *Ae. cylindrica* chromosome number and contained marker DP9. This A genome marker is likely on a translocated chromosome fragment, but may be on an A genome substitution chromosome. If it is on a translocated chromosome fragment, it may be inherited in a stable way to the next generations. All three mechanisms (recombination, translocation and chromosome retention) are well known by plant breeders who do the opposite; i.e., introgress genetic material from wild species to wheat, or produce alien addition or substitution lines to study the effect of genes of certain chromosomes or for mapping purposes (Sears 1969; Gale and Miller 1987). Although these processes usually necessitate human intervention such as manual hybridisation, post pollination hormone application or embryo rescue (Sharma 1995), some of the alien substitution lines have arisen spontaneously following hybridisation of wheat and the foreign species (Sears 1969).

It is well known that the diploid behaviour of wheat, i.e. the formation of 21 bivalents at meiosis, is controlled by the *Ph1* locus located on the long arm of chromosome 5B; if chromosome 5B is lacking, multivalents are observed at meiosis (Okamoto 1957; Riley and Chapman 1958). Therefore, in a BC1 hybrid carrying some extra A and B genome chromosomes, but not chromosome 5B, homeologous recombination may take place resulting in translocated segments. Mutants in or deletions of *Ph1* have been used to introgress genetic material into wheat by providing maximum recombination between wheat and chromosomes of alien species (e.g. Lukaszewski 2000). Furthermore, in transformed wheat using microprojectile bombardment, which is the most widely used system for wheat transformation (Janakiraman et al. 2002), transgene integration sites were often telo- or sub telomeric (Jackson et al. 2001) and thus intrinsically more prone to recombination and translocation. The same is true for transgenic barley (Salvo-Garrido et al. 2004). It was also shown that transgene integration site could correlate with chromosomal rearrangements in hexaploid *Avena sativa* L. (Svitashev et al. 2000).

The probability of transgene flow from the A and B genome of wheat to *Ae. cylindrica* is certainly reduced compared to the D genome; the extent of this reduction is unknown and difficult to hypothesise. Similarly, in allotetraploid *Brassica napus* L. (AACC genome), transgene flow to diploid *Brassica rapa* L. (AA genome) is lowered if the transgene is located on the unshared C genome, but it cannot be avoided because of intergenomic recombination between homeologous A and C genome chromosomes (Metz et al. 1997; Stewart et al. 2003). Furthermore, if a transgene confers a selective advantage to the hybrid and the subsequent introgressants, it is likely that it would be rapidly integrated into *Ae. cylindrica* on whatever genome it is placed, specially in a context of extensive hybridisation in winter wheat fields invaded by the wild species, as was observed in Oregon, USA (Morrison et al. 2002a; Morrison et al. 2002b). There are probably some exceptions to this as it was observed that in different wheat lines some marker loci or chromosomes, like chromosome 2B, display segregation distortion in favour of one parental line, lowering the rate of recombination between loci (Paillard et al. 2003). It is hypothesised that segregation distortion in this chromosome involves a genetic factor. If a transgene is flanked by a particularly inefficient distorter allele, one would expect a low segregation ratio of the further. One way to identify A or B genome chromosomes or chromosome regions that have no or only reduced ability of introgression would be to construct an interspecific linkage map between wheat and *Ae. cylindrica*.

Because introgression of wheat A and B genome fragments into *Ae. cylindrica* can occur, we conclude that it is not sufficient to place a transgene on those genomes to efficiently avoid its introgression. The risk that a transgene located on the A or B genome of wheat is maintained in an *Ae. cylindrica* background depends finally on the frequency of chromosome retentions and translocations for each genomic and chromosomal region of wheat. The markers described in this study provide tools for further investigations that are needed to obtain a clear insight into the mechanisms influencing these two factors.

Acknowledgments

We are grateful to Prof. Beat Keller and Dr. Nabila Yahiaoui of the Institute of Plant Biology of the University of Zürich, Switzerland, for providing the nulli-tetrasomic wheat lines; to Dr. Eva Stoger of the Institute for Biology of the RWTH Aachen, Germany, for providing seeds of the wheat variety Bobwhite;

to Dr. Yong-Ming Yuan and Dr. Philippe Chassot from our laboratory for advice during sequencing, to Anouk Béguin and Sarah Mamie for technical assistance, and to two anonymous reviewers for their helpful comments. This project was funded by the National Centre of Competence in Research (NCCR) Plant Survival, a research programme of the Swiss National Science Foundation.

References

Abranches R, Santos AP, Wegel E, Williams S, Castilho A, Christou P, Shaw P, Stoger E (2000) Widely separated multiple transgene integration sites in wheat chromosomes are brought together at interphase. Plant J. 24:713-723

Badaeva ED, Amosova AV, Muravenko OV, Samatadze TE, Chikida NN, Zelenin AV, Friebe B, Gill BS (2002) Genome differentiation in *Aegilops*. 3. Evolution of the D-genome cluster. Plant Syst Evol 231:163-190

Cadalen T, Boeuf C, Bernard S, Bernard M (1997) An intervarietal molecular marker map in *Triticum aestivum* L. em Thell and comparison with a map from a wide cross. Theor Appl Genet 94:367-377

Devos KM, Gale MD (1992) The use of random amplified polymorphic DNA markers in wheat. Theor Appl Genet 84:567-572

Donald WW, Ogg AG (1991) Biology and control of jointed goatgrass (*Aegilops cylindrica*), a review. Weed Technol 5:3-17

Ellstrand NC (2003) Dangerous liaisons? When cultivated plants mate with their wild relatives. In: SM Scheiner (ed) Syntheses in ecology and evolution. The Johns Hopkins University Press, Baltimore, pp 26-49

Gale MD, Miller TE (1987) The introduction of alien genetic variation in wheat. In: FGH Lupton (ed) Wheat breeding. Its scientific bases. Chapman and Hall, London New York, pp 173-210

Guadagnuolo R, Savova-Bianchi D, Felber F (2001) Gene flow from wheat (*Triticum aestivum* L.) to jointed goatgrass (*Aegilops cylindrica* Host.), as revealed by RAPD and microsatellite markers. Theor Appl Genet 103:1-8

Hegde SG, Waines JG (2004) Hybridization and introgression between bread wheat and wild and weedy relatives in North America. Crop Sci 44:1145-1155

Hernandez P, Martin A, Dorado G (1999) Development of SCARs by direct sequencing of RAPD products: a practical tool for the introgression and marker-assisted selection of wheat. Mol Breeding 5:245-253

Hernandez P, de la Rosa R, Rallo L, Dorado G, Martin A (2001) Development of SCAR markers in olive (*Olea europaea*) by direct sequencing of RAPD products: applications in olive germplasm evaluation and mapping. Theor Appl Genet 103:788-791

Jackson SA, Zhang P, Chen WP, Phillips RL, Friebe B, Muthukrishnan S, Gill BS (2001) High-resolution structural analysis of biolistic transgene integration into the genome of wheat. Theor Appl Genet 103:56-62

Janakiraman V, Steinau M, McCoy SB, Trick HN (2002) Recent advances in wheat transformation. In Vitro Cell Dev Biol-Plant 38:404-414

Kroiss LJ, Tempalli P, Hansen JL, Vales MI, Riera-Lizarazu O, Zemetra RS, Mallory-Smith CA (2004) Marker-assessed retention of wheat chromatin in wheat (*Triticum aestivum*) by jointed goatgrass (*Aegilops cylindrica*) backcross derivatives. Crop Sci 44:1429-1433

Lin Y (2001) Risk assessment of *bar* gene transfer from B and D genomes of transformed wheat (*Triticum aestivum*) lines to Jointed goatgrass (*Aegilops cylindrica*). Journal of Anhui Agricultural University 28:115-118

Lukaszewski AJ (2000) Manipulation of the 1RS.1BL translocation in wheat by induced homoeologous recombination. Crop Sci 40:216-225

Metz PLJ, Jacobsen E, Nap JP, Pereira A, Stiekema WJ (1997) The impact on biosafety of the phosphinothricin-tolerance transgene in inter-specific *B. rapa* x *B. napus* hybrids and their successive backcrosses. Theor Appl Genet 95:442-450

Morrison LA, Crémieux L, Mallory-Smith CA (2002a) Infestations of jointed goatgrass (*Aegilops cylindrica*) and its hybrids in Oregon wheat fields. Weed Sci 50:737-747

Morrison LA, Riera-Lizarazu O, Crémieux L, Mallory-Smith CA (2002b) Jointed goatgrass (*Aegilops cylindrica* Host) x wheat (*Triticum aestivum* L.) hybrids: hybridization dynamics in Oregon wheat fields. Crop Sci 42:1863-1872

Nelson JC, Sorrells ME, Vandeynze AE, Lu YH, Atkinson M, Bernard M, Leroy P, Faris JD, Anderson JA (1995) Molecular mapping of wheat - major genes and rearrangements in homoeologous group-4, group-5, and group-7. Genetics 141:721-731

Okamoto M (1957) Asynaptic effect of chromosome V. Wheat Inf Ser 5:6

Paillard S, Schnurbusch T, Winzeler M, Messmer M, Sourdille P, Abderhalden O, Keller B, Schachermayr G (2003) An integrative genetic linkage map of winter wheat (*Triticum aestivum* L.). Theor Appl Genet 107:1235-1242

Paran I, Michelmore RW (1993) Development of reliable PCR-based markers linked to downy mildew resistance genes in Lettuce. Theor Appl Genet 85:985-993

Rajhathy T (1960) Continuous spontaneous crosses between *Aegilops cylindrica* and *Triticum aestivum*. Wheat Inf Ser 11:20

Riley R, Chapman V (1958) Genetic control of cytologically diploid behaviour of hexaploid wheat. Nature 182:713-715

Salvo-Garrido H, Travella S, Bilham LJ, Harwood WA, Snape JW (2004) The distribution of transgene insertion sites in barley determined by physical and genetic mapping. Genetics 167:1371-1379

Savova D, Rufener-AlMazyad P, Felber F (1996) Cytogeography of *Medicago falcata* L. and *M. sativa* L. in Switzerland. Bot Helv 106:197-207

Sears ER (1966) Nullisomic-tetrasomic combinations in hexaploid wheat. In: Riley R Lewis KR (eds) Chromosome manipulations and plant genetics. Oliver and Boyd, London pp 29-45

Sears ER (1969) Wheat cytogenetics. Annu Rev Genet 3:451-468

Sharma HC (1995) How wide can a wide cross be? Euphytica 82:43-64

Slageren MW van (1994). Wild wheats: a monograph of *Aegilops* L. and *Amblyopyrum* (Jaub. & Spach) Eig (Poaceae). Wageningen Agricultural University Press, Wageningen; ICARDA, Aleppo

Stewart CN, Halfhill MD, Warwick SI (2003) Transgene introgression from genetically modified crops to their wild relatives. Nat Rev Genet 4:806-817

Svitashev S, Ananiev E, Pawlowski WP, Somers DA (2000) Association of transgene insertion sites with chromosome rearrangements in hexaploid oat. Theor Appl Genet 100:872-880

Wang ZN, Hang A, Hansen J, Burton C, Mallory-Smith CA, Zemetra RS (2000) Visualization of A- and B-genome chromosomes in wheat (*Triticum aestivum* L.) x jointed goatgrass (*Aegilops cylindrica* Host) backcross progenies. Genome 43:1038-1044

Zemetra RS, Hansen J, Mallory-Smith CA (1998) Potential for gene transfer between wheat (*Triticum aestivum*) and jointed goatgrass (*Aegilops cylindrica*). Weed Sci 46:313-317

CHAPTER 3

Molecular analysis, cytogenetics and fertility of introgression lines from transgenic wheat to *Aegilops cylindrica* Host[3]

---
[3] *Published in Genetics 174: 2061-2070 (December 2006). DOI: 10.1534/genetics.106.058529*

Abstract

Natural hybridization and backcrossing between *Aegilops cylindrica* and *Triticum aestivum* can lead to introgression of wheat DNA into the wild species. Hybrids between *Ae. cylindrica* and wheat lines bearing herbicide (*bar*), reporter (*gus*), fungal disease resistance (*kp4*), and increased insect tolerance (*gna*) transgenes were produced by pollination of emasculated *Ae. cylindrica* plants. F1 hybrids were backcrossed to *Ae. cylindrica* under open pollination conditions, and first backcrosses were selfed using pollen bags. Female fertility of F1 ranged from 0.03% to 0.6%. Eighteen percent of the sown BC1s germinated and flowered. Chromosome numbers ranged from 30 to 84 and several of the plants beared wheat-specific sequence characterized amplified regions (SCARs) and the *bar* gene. Self-fertility in two BC1 plants was 0.16% and 5.21%, the others were completely self-sterile. Among 19 BC1S1 individuals one plant was transgenic, had 43 chromosomes, contained the *bar* gene and survived glufosinate treatments. The other BC1S1 plants had between 28 and 31 chromosomes, several of them carried SCARs specific to wheat A and D genomes. Fertility of these plants was higher under open pollination conditions than by selfing and did not necessarily correlate with even or euploid chromosome number. Some individuals having supernumerary wheat chromosomes recovered full fertility.

Introduction

Aegilops cylindrica Host is a widespread Mediterranean, Western Asiatic and even circumboreal element, present as an adventive in Northern Italy, France, Switzerland, and other countries of Western, Northern and Eastern Europe (VAN SLAGEREN 1994). Probably introduced in the USA at the end of the 19th century, *Ae. cylindrica* is now widespread there and is considered as a troublesome weed in wheat fields (DONALD and OGG 1991; VAN SLAGEREN 1994). Indeed, *Ae. cylindrica* is one of the species in the genus showing most pronounced tendency to weediness (VAN SLAGEREN 1994). Increased weediness due to transgene introgression into wild populations of *Ae. cylindrica*, for example in the case of herbicide resistance genes, is a major concern about the cultivation of genetically modified wheat (HEGDE and WAINES 2004).

Allotetraploid *Ae. cylindrica* (2n=4x=28, DDCC genome) and allohexaploid bread wheat (*Triticum aestivum* L., 2n=6x=42, BBAADD genome) (WAINES AND BARNHART 1992) form natural hybrids where they grow in sympatry (MORRISON *et al.* 2002a; RAJHATHY 1960; VAN SLAGEREN 1994). Morphologically, F1 hybrids are completely uniform and appear to be intermediate between the parental species, although some parental characters, like glume stiffness and hairiness are dominantly expressed in the hybrids (BELEA 1968). This uniformity contrasts with the morphological variation of the offspring of the hybrids (BC1) for spike shape, leaf shape and growth habit, characters that can't be used to identify the pollen parent (SNYDER et al. 2000).

Wheat and *Ae. cylindrica* share the D genome (KIMBER and ZHAO 1983), issued from their common diploid ancestor *Aegilops tauschii* Coss. First generation hybrids always have 35 chromosomes (RAJHATHY 1960). At meiosis in pentaploid hybrids (2n=5x=35, DDCBA genome), the D genome chromosomes of wheat pair with those of *Ae. cylindrica* while A, B and C genome chromosomes remain as univalents (ZEMETRA et al. 1998). The highly disrupted meiosis in F1 hybrids results in complete male sterility and extremely reduced female fertility, determined as the percentage of florets bearing seeds (WANG et al. 2001; ZEMETRA et al. 1998).

In field grown *T. aestivum* x *Ae. cylindrica* F1 hybrids, female fertility was 2.2% when backcrossed with *Ae. cylindrica* (ZEMETRA et al. 1998), whereas in another study the same cross resulted in 0.87% female fertility (WANG et al. 2001). Average percent seed set in *T. aestivum* x *Ae. cylindrica* F1 hybrids, when surrounded by different proportions of the parental species, was 2.3% and 3.8% in two different years while self fertility of BC1s averaged 0.3% and 0.06% in two years (SNYDER et al. 2000). In Oregon, USA, female fertility of hybrids collected in the field was 1% but the recurrent parent was not established (MORRISON et al. 2002a).

Female fertility in BC1s, when pollinated by *Ae. cylindrica* was 4.4%, at each successive backcross to *Ae. cylindrica* mean fertility increases, the BC2 (second generation backcross) already had 6.9% self fertility and 18% female fertility, whereas in the BC2S2 (second backcross selfed twice) generation, self fertility was restored to 78,9% (WANG et al. 2001). Fertility in these backcrosses varied quite substantially among individuals, often ranging from 0% seed set to a high proportion of fertile florets.

The first backcross generation of hybrids, bearing always 35 chromosomes (RAJHATHY 1960), is variable in chromosome numbers. In BC1 plants issued from *T. aestivum* x *Ae. cylindrica* crosses, with *Ae. cylindrica* as the recurrent parent, chromosome numbers determined upon 15 plants ranged from 30 to 49 chromosomes (ZEMETRA et al. 1998), in another study the same type of BC1 plants had 34 to 49 chromosomes (based on 10 individuals), and an average of 12 C genome chromosomes (issued from *Ae. cylindrica*) (WANG et al. 2002).

Experimental research in the USA involved essentially *T. aestivum* x *Ae. cylindrica* hybrids and their backcross derivatives (reviewed in HEGDE and WAINES 2004). However, it is supposed that when spontaneous hybridization takes place, the wild plant generally acts as the female parent, leading to unidirectional gene flow (LADIZINSKY 1985). *Ae. cylindrica* x *T. aestivum* hybridization would be the preferential way for crop-to-wild gene flow to occur. BC1 plants where *Ae. cylindrica* was the female parent in F1 hybridization and the male in the backcross had only 28 to 29 chromosomes, an *Ae. cylindrica* like morphology and restored fertility (GUADAGNUOLO et al. 2001).

By the use of GISH technology, it was demonstrated that introgression of wheat DNA from the A or B genomes into *Ae. cylindrica* was possible starting from F1 hybrids where wheat acted as the female parent; BC2S2 plants had up to three extra, non homologous A or B genome chromosomes (WANG et al. 2000). In an introgressive series where wheat acted as the male progenitor, DNA markers specific to wheat A and D genome chromosomes were detected in BC1S1 plants having euploid *Ae. cylindrica* chromosome number (2n=28) or up to three additional chromosomes (SCHOENENBERGER et al. 2005). The analysis of the pattern of retention of D genome chromosomes in *T. aestivum* x *Ae. cylindrica* backcross derivatives, using wheat D genome-specific microsatellite markers, has shown that most alleles of the D genome of wheat are neither preferentially inherited nor lost and may be retained in a *Ae. cylindrica* population (KROISS et al. 2004).

Wheat genes may be expressed in *Ae. cylindrica* when introgressed. Thus, imidazolinone resistant wheat having the resistance gene on chromosome 6D (ANDERSON et al. 2004) was used to produce *T. aestivum* x *Ae. cylindrica* hybrids and BC1 to either of the progenitors, six of the seven recovered BC1 plants were resistant to imazamox treatment (SEEFELDT et al. 1998). Moreover, it was shown in a reduced sample of experimentally produced *Ae. cylindrica* x *T. aestivum* hybrids and BC1s with *Ae. cylindrica* as the pollen parent, that a herbicide resistance trait conferred by the *bar* gene, was inherited to the BC1s only if located on the D genome of the transformed wheat progenitors, whereas if it was located on the B genome none of the BC1s were resistant against glufosinate (LIN 2001).

The aim of the present work was to simulate a natural introgressive series from transgenic wheat to *Ae. cylindrica* with the wild species as the female parent, and to assess, at each generation, parameters which are essential to risk assessment like fertility, survival, chromosome constitution, introgression of wheat-specific sequence characterized amplified regions (SCARs) and transgenes and their expression.

Materials and methods

Plant material: Transgenic *Aegilops cylindrica* Host x *Triticum aestivum* L. hybrids, first backcross generations (BC1) with *Ae. cylindrica* as the recurrent parent, and selfed offspring of the first backcrosses (BC1S1) were generated in a greenhouse. Hybrids produced by crosses with near-isogenic wheat varieties (wild type, WT), were grown in outdoor beds. Numbering of the individuals follows the genealogy of the plants, F1 Hybrids producing viable seeds were numbered from 1 to 13, as well as their BC1 offspring (e.g. F1-1 is the mother plant of BC1-1, etc.). All plants were potted in 9 cm square pots, except of the BC1, which were planted in pots of 20 cm diameter, in order to recover bigger plants. All *Ae. cylindrica* plants originated from seeds collected from distinct individuals in a natural population in Sierre, Switzerland (GUADAGNUOLO *et al.* 2001). All plants were sown in autumn, except of the spring wheat varieties used as pollen donors. Several transgenic spring wheat lines and their near-isogenic counterparts were used as pollen donor for hybridizations (Table 1).

Cytogenetic analysis: Chromosomes were counted at mitosis in root tip meristemes, and the number of observed satellitiferous chromosomes was noted. Root tips were pre-treated in a saturated water solution of α-bromonaphtalene for 180 min at room temperature. Meioses were observed from pollen mother cells in anthers. Root tips and immature anthers were fixed for at least one week in absolute ethanol and glacial acetic acid (3:1) added with acetocarmine and traces of iron acetate. Before staining, root tips were treated in a water solution containing 5% pectinase and 2% cellulase (w/v) for softening of the tissue. After fixation, root tips or anthers were stained in acetocarmine (1%) added with traces of iron acetate for an hour, then heated gently for 2 min over a flame. The material was then stored in 45% acetic acid, squashed and observed at 1000x magnification under a light microscope.

Table 1. Wheat lines used for hybridizations, number of hybrids produced and sown, germination rate, survival rate to flowering and number of spikes produced

pollen donor variety / transformation event	transgene	chromosome location	hybrid seeds produced	hybrids sown	germination rate (%)	survival to flower (%)	mean number of spikes/individual
Greina 16[a]	KP4 bar	-	248	150	80	71.3	2.7
Golin 5[a]	KP4 bar	-	353	150	88	80	4.3
Bobwhite 1[b]	gus bar	2A	3	0	-		
Bobwhite 3[b]	gus bar	6B	28	0	-		
Bobwhite 5[b]	bar	1D	27	21	90.5	90.5	3.1
Bobwhite 7[b]	gus bar	2B	13	9	88.8	88.8	2.8
Bobwhite 8[b]	gus bar	6A	38	24	83.3	83.3	2.7
Bobwhite 92[c]	gna	-	14	0	-		
Bobwhite 99[c]	gna	-	90	22	86.4	86.4	2.5
Bobwhite 150[c]	gna	-	12	0	-		
Greina (wild type)	-		89	48	93.8	83.3	8.8
Golin (wild type)	-		56	27	100	100	6.2
Bobwhite (wild type)	-		88	0	-		

[a] (Clausen et al. 2000) [b] (Abranches et al. 2000) [c] (Stoger et al. 1999)

Greenhouse experiments: Ae. cylindrica plants were emasculated, spikes covered with pollen-proof bags, pollinated with wheat pollen two to five days after, and bagged again during maturation of the seeds. Transgenic F1 hybrids were grown in a greenhouse surrounded by a dense planting of Ae. cylindrica. Ventilators were set around the artificial plot during flowering for maximizing pollen flow from Ae. cylindrica to the hybrids. The BC1 seeds produced were all sown in a greenhouse. In order to evaluate self fertility all spikes were bagged. The BC1S1s were sown in pots for further analysis, up to 15 spikes per plant were covered with pollen bags in order to evaluate self-fertility, the remaining spikes were exposed to pollen of a surrounding Ae. cylindrica population of at least 2000 plants in order to evaluate fertility under open pollination conditions.

Fertility assessment: Hybrids between wheat and *Ae. cylindrica* are completely male sterile (WANG *et al.* 2001), only female fertility was assessed. Female fertility was calculated as follows: number of flowers that produced a seed/total number of flowers x 100. Similarly, self-fertility of the BC1 was indicated as the number of flowers that produced a seed/total number of flowers x 100. Spikelets are axes bearing an indeterminate number of alternate flowers, but only the basal two to three (rarely four) were completely developed, the apical ones were always aborted. In order to estimate the number of flowers produced, we divided hybrid spikes into two regions: the basal part where spikelets have three completely developed flowers, and the upper part, where the spikelets contain two well-formed flowers. Basal spikelets were bigger in size, with the lemmas of the third and higher order flower well protruding from the glumes, whereas apical spikelets were smaller and more compact. In order to determine the fertility of BC1S1 plants all flowers were counted individually. Differences between self and open pollination fertilities of BC1S1 were assessed with the Wilcoxon ranked signed test, using SPSS 11 (SPSS Inc.).

DNA extraction and PCR: Genomic DNA was extracted from a bulk of five individuals of each transgenic and their near isogenic wheat lines, and from individual hybrids, BC1s, BC1S1s and *Ae. cylindrica* of the mother population, following an SDS-Na-acetate protocol (SAVOVA-BIANCHI 1996). The *bar* gene conferring resistance to phosphinothricin herbicides, and the wheat-specific DNA marker IB10 (SCHOENENBERGER *et al.* 2005) were amplified in 132 Golin-5 hybrids and 120 Greina-16 hybrids produced. Furthermore, we performed PCRs using the different wheat-specific SCAR markers DP9, D1P9, GB10 and the *bar* gene in all individuals of the introgressive series which produced an offspring, these latter reactions were repeated at least twice. Amplification conditions of the *bar* gene followed TAKUMI and SHIMADA (1996) whereas PCR protocols and chromosome assignment of the SCAR markers were described in SCHOENENBERGER *et al.* (2005). Wheat cv. Chinese Spring (CS) and its nulli-tetrasomic (NT) aneuploids for chromosome assignment of introgressed markers (Fig. 2) were obtained from Beat Keller of the Institute of Plant Biology of the University of Zürich, Switzerland.

DNA sequencing: Both strands of the *bar* gene, amplified in the transgenic wheat pollen donor plants as well as in a BC1S1 individual, were sequenced as described by CHASSOT *et al.* (2001) in an ABI 310 automated sequencer (Applied Biosystems), with a modified cycle sequencing temperature profile: 96°C for 10 sec, 55°C for 5 sec, and 60°C for 4 min.

Recovered sequences were compared with GenBank's nr database sequences to verify homology with the *bar* gene.

Herbicide treatment: To test expression of the *bar* gene, single marked leaves of BC1S1 plants were sprayed with 150mg/l Basta® (Bayer CropScience). Two weeks after application, herbicide susceptibility was assessed by scoring necrosed leaves. The assay was repeated twice on distinct leaves.

Results

F1 hybrids: Manual *Ae. cylindrica* x *T. aestivum* hybridizations yielded a total of 1059 F1 hybrid seed, 451 of these were sown for the production of BC1 (Table 1). Hybrids issued from some of the crosses were not sown due to the low number of produced seeds. Overall germination rate of the hybrids was 86.5%, and 79.8% of the sown individuals reached maturity (i.e. produced at least one spike). Mortality of the plants was mainly due to frost during the winter, and to some rare dwarf plants which died before maturity. Being planted in pots, the hybrids produced only small amounts of tillers and spikes (Table 1). Non-transgenic hybrids were grown outside and produced more tillers, while the transgenic ones were grown indoor for legal reasons.

Ae. cylindrica x *T. aestivum* hybrids always had 35 chromosomes, corresponding to the expected pentaploid value (Table 2, Figure 1). PCRs of the wheat-specific DNA fragment IB10 and the *bar* gene in all 252 Golin 5 and Greina 16 hybrids demonstrated the hybrid nature and the presence of the transgene in the manually produced plants. Only three individuals out of 132 Golin 5 hybrids lacked the IB10 fragment and four missed the *bar* gene, whereas both fragments were detected in all Greina 16 hybrids (data not shown). Lack of amplification was attributed to the bad quality of the DNA extract rather than to loss of the fragments. Furthermore, PCRs of wheat-specific SCAR markers performed on all individuals of the introgressive series which produced an offspring, showed that all hybrids bearing viable seed possess the specific wheat markers (Table 2, Figure 2). The *bar* gene was amplified in all hybrids except one (F1 13), issued by a cross with Bobwhite 5 (Figure 3). Mean female fertility of the hybrids, was 0.2 % and varied between 0.03 % and 0.6 %. In other words, we found one BC1 seed each two to 18 hybrid individuals, depending on the wheat variety implicated in the *Ae. cylindrica* x *T. aestivum* hybridization (Table 3).

Table 2. Chromosome number, presence of satellitiferous chromosomes, wheat genome-specific SCARs (DP9, GB10, D1P9) and a transgene (*bar*) and its expression in the introgressive series

Individual	Chromosome number 2n=	No. counts	Satellitiferous chromosomes	DP9 (6A)	GB10 (5B)	D1P9 (6D)	bar gene	basta test
Ae. cylindrica	28	3	1 (1C)	-	-	-	-	+
GM-T. aestivum	42 (literature)		2 (1B & 6B)	+	+	+	+	
F1								
1	35	2		+	+	+	+	
2	35	3	1	+	+	+	+	
3				+	+	+	+	
4	35	3		+	+	+	+	
5	35	2		+	+	+	+	
6	35	2		+	+	+	+	
8	35	2		+	+	+	+	
9	35	2	1	+	+	+	+	
10				+	+	+	+	
11				+	+	+	+	
12	35	2	1	+	+	+	+	
13	35	2		+	+	+	-	
Go5-12	35	4						
Go5-136	35	2						
Go5-85	35	1						
Go5-21	35	1						
Go5-127	35	1						
Go5-128	35	1						
Go5-48	35	2						
BW 5-5	35	2						
Gr-21	35	2	1					
BC1								
1	49	4	2	+	+	+	-	
2	49	3	2	+	+	-	-	
3	36	5		+	+	-	+	
4	43-44	1		+	+	+	+	
5	49	2		+	+	+	+	
6	44	3		+	+	+	-	
7	84	1		+	+	+	+	
8	41	5	2	+	+	-	+	
9	40	5		+	-	-	+	
10	38	4		+	-	+	+	
11	41	3	2	+	-	+	+	
12	34	3		+	-	+	-	
13	30	4	≥1	+	-	+	-	

(continued)

Table 2. *Continued*

Individual	Chromosome number 2n=	No. counts	Satellitiferous chromosomes	DP9 (6A)	GB10 (5B)	D1P9 (6D)	bar gene	basta test
BC1S1								
11-1	43	7		+		+	+	-
13-1	28	3		-	-	-	-	+
13-2	29 + t	4		+	+	-	-	+
13-3	29	4		-	+	-	-	+
13-4	29	3		-	-	-	-	+
13-5	28	3		-	+	-	-	+
13-6	-			-	-	-	-	+
13-7	30	3		-	+	-	-	+
13-8	29	6		+	+	-	-	+
13-9	31	4		+	+	-	-	+
13-10	28	1		-	-	-	-	+
13-11	28	5	2	+	+	-	-	+
13-12	28	4		-	+	-	-	+
13-13	30	4		-	+	-	-	+
13-14	29	3		+	+	-	-	+
13-15	28	5		-	+	-	-	+
13-16	28	4		-	+	-	-	+
13-17	28	3		-	-	-	-	+
13-18	30	3		-	+	-	-	+

+: present/ sensitive; -: absent/ resistant; t: telocentric; Go5: Golin5, Gr: Greina WT; BW5: Bobwhite 5. A part of the information relative to the BC1S1 13 family was already published in Schoenenberger *et al.* (2005).

BC1: In total, 74 BC1 seeds were recovered, 68 of them issued from hybrids with GM wheat and six from hybrids with conventional wheat. Seventy-three of the BC1 seeds were sown (Greina 16: six seeds, Greina WT: three seeds, Golin 5: 40 seeds, Golin WT: three seeds, Bobwhite 5: nine seeds, Bobwhite 7: two seeds, Bobwhite 8: five seeds, Bobwhite 99: five seeds). Thirteen BC1 plants germinated and flowered (17.8% survival). Chromosome numbers ranged from 30 to 84 chromosomes (Table 2, Fig. 1). Male meioses in the BC1 were mostly irregular, with a high proportion of univalents at metaphase one (Table 4). We observed several bivalents and some trivalents; in one cell of a BC1 individual we observed a tetravalent. Pollen grains were mostly collapsed and irregular in size, and considered to be sterile. Only in some individuals we found a proportion of well-formed turgescent pollen grains, whose presence correlated with some self-fertility of the plant. Amplifications of wheat genome-specific SCAR markers showed the presence of wheat-DNA from A, B and D genomes in several of the BC1 plants (Table 2). The *bar* gene was maintained in eight out of 13 BC1 plants (Fig. 3). Most of the BC1 were completely self-sterile. However, two BC1 plants (BC1-11 and BC1-13) produced an offspring, self-fertility was 0.16% and 5.21% respectively (Table 3). BC1 plants displayed a high morphologic variability, traits ranged from *Ae. cylindrica* like to F1 hybrid like. Growth habit ranged from squarrose (*Ae. cylindrica* like) to erect (wheat like), leaf width ranged from 4 mm to 13 mm and spike morphology from *Ae. cylindrica* like to F1 hybrid like (data not shown). Spike emergence was not synchronous, there was a delay of 22 days between the first plant to flower and the last ones. One plant (BC1-7) grew extremely slowly and flowered 45 days after the first.

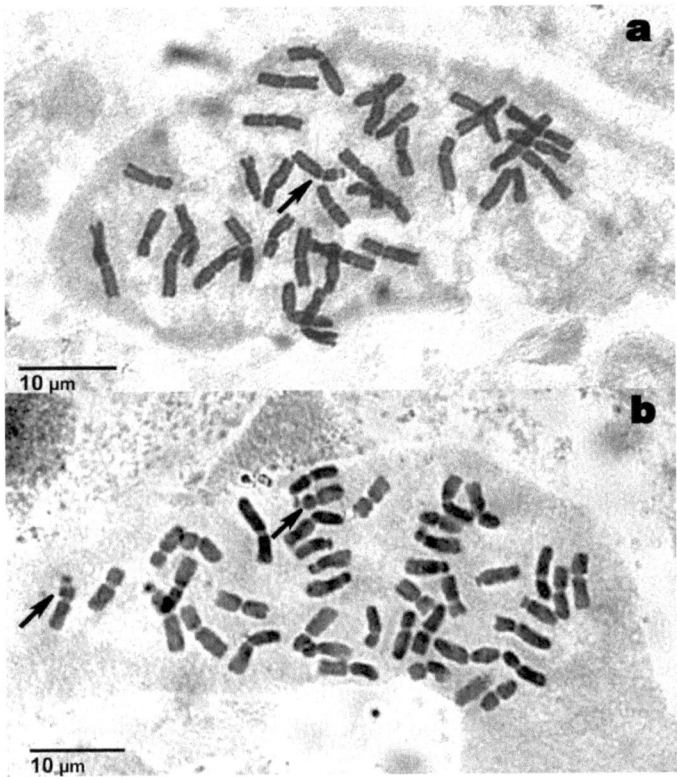

Figure 1. Metaphases in root tip cells of a: *Ae. cylindrica* x *T. aestivum* F1 hybrid with 2n=35 chromosomes, and b: BC1-11 with 2n=41 chromosomes, arrows indicate satellitiferous *Ae. cylindrica* chromosome 1C

Table 3. Fertility of *Aegilops cylindrica* x *Triticum aestivum* hybrids, BC1 and BC1S1

F1	progenitor	flowers	seeds/individuals	female fertility (%)
	Golin 5	18197	0.33	0.22
	Golin WT	4712	0.11	0.06
	Greina 16	8495	0.056	0.07
	Greina WT	11383	0.075	0.03
	Bobwhite 5	1798	0.53	0.6
	Bobwhite 7	806	0.25	0.2
	Bobwhite 8	1825	0.25	0.3
	Bobwhite 99	1634	0.26	0.3

BC1	progenitor	flowers	seeds	self fertility (%)
1	Golin 5	913	0	0
2	Golin 5	780	0	0
3	Golin 5	1756	0	0
4	Golin 5	858	0	0
5	Golin 5	987	0	0
6	Golin 5	667	0	0
7	Golin 5	82	0	0
8	Greina 16	911	0	0
9	Bobwhite 8	1051	0	0
10	Bobwhite 8	907	0	0
11	Bobwhite 8	1234	2	0.16
12	Bobwhite 5	624	0	0
13	Bobwhite 5	768	40	5.21

BC1S1	BC1 parent	flowers		seeds		fertility (%)	
		self	open	self	open	self	open
11-1	11	459	489	0	0	0	0
13-1	13	94	87	1	6	1.1	6.9
13-2	13	14	84	7	60	50	71.4
13-3	13	-	160	-	26	-	16.3
13-4	13	286	104	0	16	0	15.4
13-5	13	245	130	42	79	17.1	60.8
13-6	13	230	106	192	93	83.5	87.7
13-7	13	298	121	0	18	0	14.9
13-8	13	164	94	17	32	10.4	34
13-9	13	14	97	3	30	21.4	30.9
13-10	13	-	148	-	45	-	30.4
13-11	13	130	140	0	12	0	8.6
13-12	13	-	125	-	103	-	82.4
13-13	13	-	114	-	65	-	57
13-14	13	28	111	1	80	3.6	72.1
13-15	13	4	91	1	63	25	69.2
13-16	13	186	71	124	57	66.7	80.3
13-17	13	176	92	116	69	65.9	75
13-18	13	-	98	-	75	-	76.5

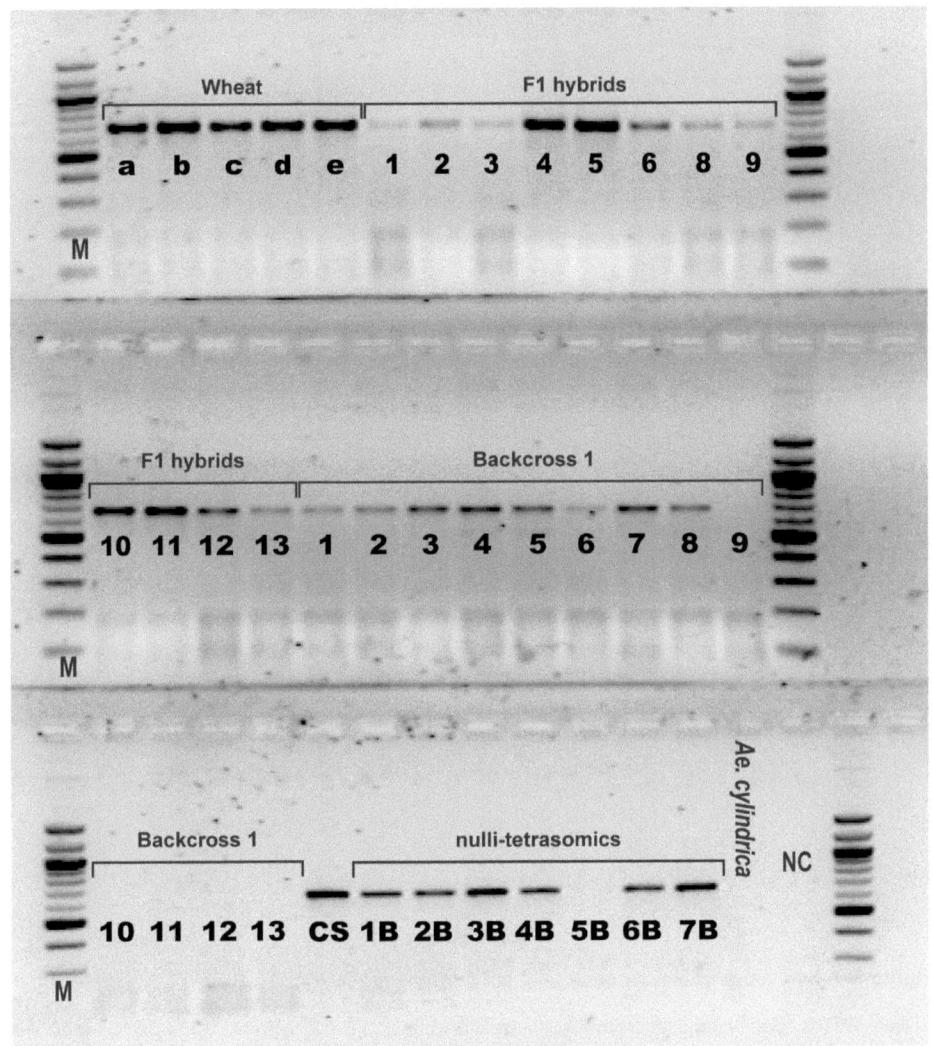

Figure 2. Amplification of the GB10 SCAR marker (662 bp) in wheat (a: Golin 5, b: Greina 16, c: Bobwhite WT, d: Bobwhite 5, e: Bobwhite 8), F1 hybrids, BC1, CS-NT lines and *Ae. cylindrica*. NC: negative control, M: 100 bp ladder.

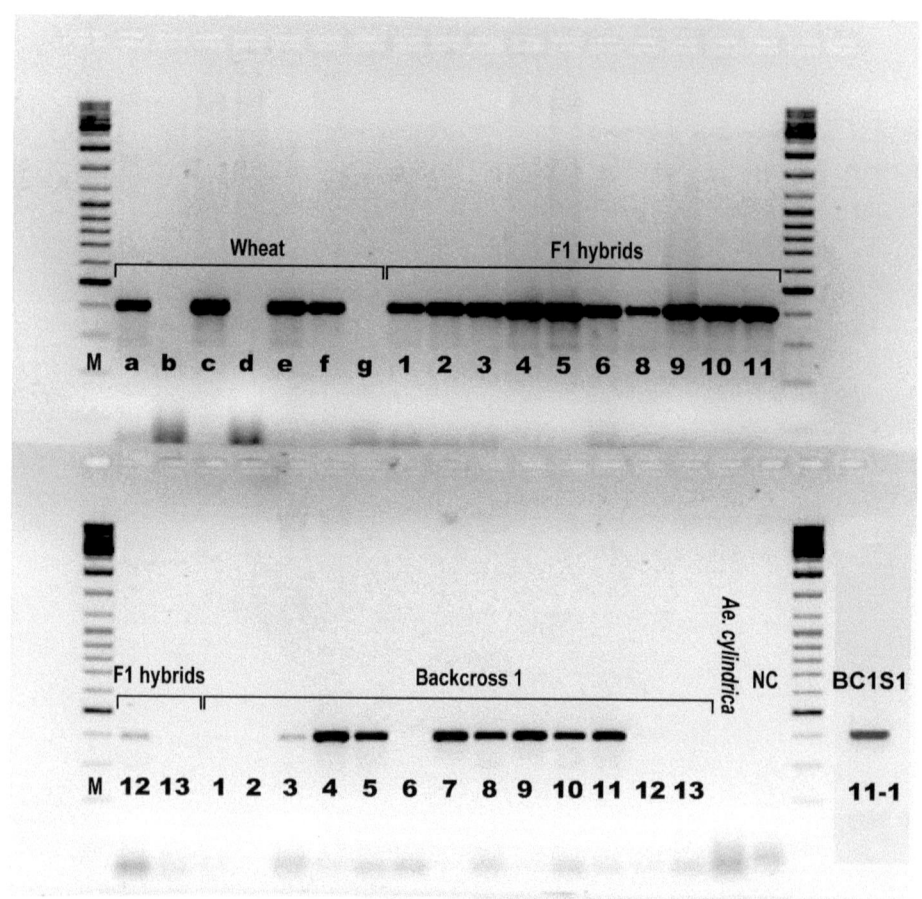

Figure 3. Amplification of the *bar* gene (402 bp) in wheat (a: Golin 5, b: Golin WT, c: Greina 16, d: Greina WT, e: Bobwhite 8, f: Bobwhite 5, g: Bobwhite WT), F1 hybrids, BC1, BC1S1 and *Ae. cylindrica*. NC: negative control, M: 100 bp ladder.

Table 4. Meiosis in BC1

BC1 No.	2n	meiotic configuration (approximate)				micro-nuclei	collapsed pollen	regularity of pollen diameter
		I	II	III	IV			
1	49					-	-	-
2	49	+	+	-	=	=	- some very big
3	36	-	+			-	+	-
4	43-44						+	--
5	49	49 in some cells	=	-		+	+	-
6	44	+	=			-	-	=
7	84					+	+	--
8	41	+	-, in some cells +			=	-	-
9	40	-	+			=	=	-
10	38	12	13	-		-	=	-
11	41	=	=			=	=	-
12	34	=	=	-	1	-	-	-
13	30	-	+			-	-	=

-: few; =: some; +: many; regularity of pollen diameter: --: very irregular; -: irregular; =: quite regular

BC1S1: A total of 42 BC1S1 seeds were recovered from two BC1 plants (Table 3) and 22 of them were sown (2 seeds from individual BC1-11 and 20 from BC1-13). One BC1S1-11 and 18 BC1S1-13 germinated and grew to maturity. BC1S1-11-1 had 43 chromosomes while chromosome numbers of the BC1S1-13 family ranged from 28 to 31 (Table 2). Wheat-specific SCAR markers DP9 and D1P9 were detected in several of the BC1S1 individuals whereas wheat-chromosome 5B specific SCAR GB10 was missing already at the BC1 generation (Fig. 2). The *bar* gene was absent in individual BC1-13 and in its entire offspring, the BC1S1-13 family. However, *bar* was present in individual BC1-11 and in its only survived descendant, BC1S1 11-1 (Fig. 3). Presence of the *bar* gene correlated with resistance to Basta® treatment, whereas all plants missing the bar gene were susceptible to the herbicide application. Sequencing of the amplification product in individual BC1S1 11-1 and in its wheat progenitor Bobwhite 8, revealed 100% identity of the sequenced 402 bp fragment between the two. Database comparison of the obtained sequence in GenBank's nr databases, using the Basic Local Alignment Search Tool (BLAST), revealed 100% sequence homology with the *Streptomyces hygroscopicus bar* gene for phosphinothricin acetyl transferase (bases 117 to 518) (WHITE et al. 1990). BC1S1 plants displayed a high range in fertility under self- and open-pollination conditions. Spikes in open pollination with pollen from Ae. cylindrica produced significantly ($p = 0.001$) more seeds than those of the self-pollination treatment. Fertility among the 19 BC1S1 individuals analyzed varied considerably; it ranged from 0% (self- and open-pollinated) to almost fully fertile (Table 3). Interestingly, fertility does not always correlate with even or euploid Ae. cylindrica chromosome number, some individuals having 28 chromosomes being much more sterile than others having 29 or even 31 chromosomes.

Discussion

Germination and survival: Germination rate of F1 hybrids is comparable with the germination rate of pure Ae. cylindrica, which was often found to be around 90% in our past experiences (data not shown). However, germination of BC1 seeds was much lower. In fact, most of the seeds sown were shriveled and did not germinate, as already observed by RAJHATHY (1960) and SNYDER et al. (2000) on reciprocal crosses. At the next generation (BC1S1), germination was again similar to pure Ae. cylindrica. The negative fitness effect of hybridization on germinability at the BC1 generation seems to be overcome at the BC1S1 generation.

Cytology: First generation hybrids always have 35 chromosomes and a DDCBA genome, whereas chromosome constitutions of BC1 plants vary considerably. Chromosome numbers of female gametes of F1 hybrids have never been established so far, but if we assume that the chromosomes of the A, B and C genomes, which are present as haploid sets (ZEMETRA et al. 1998), are randomly distributed during meiosis to the megaspores or lost, a female gamete of a hybrid should hypothetically have 17-18 chromosomes, and a BC1 plant 31-32 chromosomes. However, a female gamete of a hybrid that can be successfully fertilized by a male gamete of Ae. cylindrica seems to have usually more than 17-18 chromosomes. In fact, we observed that several BC1 plants had a high chromosome number, and assuming that Ae. cylindrica contributed with gametes of 14 chromosomes to the constitution of the BC1s, it seems that viable female gametes of F1 hybrids, which chromosome number is often between 20 and 35 maintain a high number of wheat chromosomes. BC1s having as much as 49 chromosomes were observed several times on reciprocal crosses (WANG et al. 2002; ZEMETRA et al. 1998; this study, Table 3). The most plausible explanation would be the fecundation of an unreduced female gamete of an F1 hybrid ($n=5x=35$) with a normal Ae. cylindrica male gamete, resulting in heptaploid individuals with DDDCCBA genomes. Moreover, BC1-7 was probably produced by the fecundation of a gamete of the F1 having 70 chromosomes (n=10x), resulting in a dodecaploid individual having the putative DDDDDCCCBBAA genome. DAVID et al. (2004) observed similarly that fertile *Aegilops ovata* x *Triticum turgidum durum* hybrids always produced an offspring having amphiploid chromosome number or close to it. Hybrid sterility was thus overcome by producing unreduced gametes. Moreover, the ability to produce unreduced gametes depended on the *Aegilops ovata* progenitor population. In *Ae. cylindrica* x *T. aestivum* hybrids too, high chromosome number of the BC1 indicates that fertility is restored by the production of unreduced gametes. However, we never observed self-fertile hybrids and we never counted 2n=70 (amphiploid value) in a descendant of a hybrid but we cannot exclude *a priori* that decaploid *Ae. cylindrica* x *T. aestivum* amphiploids can be produced. However, the three BC1 individuals resulting from unreduced female gametes (*i.e.* those possessing 49 chromosomes) were completely self sterile and BC1-7 grew extremely slowly, probably because of the excessive chromosome load. Under open pollination conditions, BC1 plants produced from *Ae. cylindrica* x *T. aestivum* and backcrosses had 28-29 chromosomes and restored (but not quantified) fertility (GUADAGNUOLO et al. 2001). These contrasting results may be explained by cold temperatures registered when meiosis occurred in the F1 hybrids in spring 2003 inducing meiotic abnormalities leading to enhanced frequency of unreduced gamete formation. On the other hand, we cannot exclude a genotype effect of the wheat progenitor on unreduced gamete formation.

Meiotic irregularity in BC1s is not surprising if we consider that a maximum of 14 bivalents may be formed in these individuals represented by the C and D genome chromosomes. The A and B genome chromosomes would form univalents, with the exception of individuals produced by 10x gametes where all chromosomes are doubled. In individuals originated by unreduced gametes, chromosomes of the D genome may form multivalents. Wheat A and B genome chromosomes as well as *Ae. cylindrica* C genome chromosomes may be implicated in homeologous pairing, resulting in translocated chromosome segments. Moreover, the formation of micronuclei by univalent chromosomes lost in the cytoplasm during anaphase was observed in all analyzed meioses, these micronuclei probably contain wheat A and B genome chromosomes, and are eliminated.

BC1S1 individuals with 28 chromosomes, that is euploid chromosome number, may have a complete DDCC genome and fully restored fertility (e.g. BC1S1 13-17). These individuals might have some wheat D genome DNA left (e.g. BC1S1 13-16). On the other hand, plants with 28 chromosomes may have alien substitution chromosomes, resulting in reduced (self-) fertility (e.g. BC1S1 13-1), or intergenomic translocations between wheat and *Ae. cylindrica* chromosome segments (see also SCHOENENBERGER *et al.* 2005). Plants with 29-31 chromosomes may contain wheat chromosomes present as monosomic additions or as homolog pairs, which seems plausible particularly in individuals showing a relatively high self-fertility (e.g. BC1S1 13-2 and 13-9).

Satellitiferous chromosomes are often easy to recognize, and could serve as cytological markers to identify introgressed chromosomes. Wheat chromosomes 1B and 6B have secondary constrictions (GILL *et al.* 1991), whereas in *Ae. cylindrica* a pair of satellites attached to the shorter arm of a heterobrachial chromosome is easily recognized (PRIADCENCU *et al.* 1967), and identified as chromosome 1C (LINC *et al.* 1999). Chromosome 1C was clearly recognizable, present in one copy in the F1 hybrid and as a pair in the BC1 (Table 2, Fig 1). Unfortunately, wheat chromosomes 1B and 6B were not clearly distinguishable by their secondary constrictions in the F1 hybrids and descendants, and we could not tell if they introgressed into *Ae. cylindrica*.

Crop-to-wild gene flow may occur through recombination in homologous chromosome pairs, translocation or chromosome retention (WANG *et al.* 2000, SCHOENENBERGER *et al.* 2005). Additional retained chromosomes could be stable over generations if present in homologous pairs, which may occur once male fertility restored, and aneuploid *Ae. cylindrica* populations may be founded. A fourth possible mechanism for gene flow to occur is the formation of amphiploids via unreduced gametes, enhancing the fertility of the otherwise mostly sterile hybrids (DAVID *et al.* 2004). Chromosome retention in introgressed aneuploids, may lead to production of a new, fully fertile *Aegilops* taxon, containing several additional chromosome pairs, from the A and B genomes of wheat, if a certain reproductive isolation is achieved. In fact, hybridization can lead to the formation of new taxa if intersterility between a neopolyploid or a chromosomically rearranged homoploid, and its parental species is achieved (ABBOTT 1992; ELLSTRAND and SCHIERENBECK 2000). On the other hand these polyploids can serve as a bridge for gene flow between the parental taxa (PETIT *et al.* 1999).

SCAR and *bar* gene introgression: Segregation was observed in the BC1 generation for marker D1P9 and GB10. Surprisingly, marker GB10 was maintained in all BC1s having a wheat progenitor of the variety Golin and Greina, whereas, no BC1 maintained this SCAR if Bobwhite was the progenitor. Marker DP9 was maintained in all BC1 although it is specific to the A genome of wheat. Apparently different wheat chromosomes or chromosome regions have different abilities to be maintained as backcrossing takes place.

Furthermore, there is a varietal effect of the wheat progenitor on segregation of its chromosomes or chromosome fractions in an *Ae. cylindrica* background. Similarly, in wheat intervarietal linkage maps, segregation distortion in favour of one of the parental varieties was often observed (e.g. CADALEN *et al.* 1997; PAILLARD *et al.* 2003). D genome marker D1P9 segregated at a higher than expected ratio of 1:1 in the BC1. In fact, upon 13 BC1 plants, nine had the wheat-specific marker D1P9. Deviation from expected segregation of microsatellite markers specific to the D genome of wheat was observed in *T. aestivum* x *Ae. cylindrica* BC1s having *Ae. cylindrica* as the recurrent parent, interestingly deviation was always due to a higher ratio of wheat alleles than expected (KROISS *et al.* 2004). However, these alleles were not located on chromosome 6D as in our case, and the wheat varieties used were not the same.

BC1 individual 13 had both SCARs DP9 and D1P9 in a hemizygous state. Its selfed offspring segregated for marker D1P9, in fact on 18 plants 13 contained the marker (3:1 ratio). Marker DP9 is on an A genome chromosome, it is present in a hemizygous state in all BC1, and it was passed to the BC1-13 family either as an added monosome, as a substitution chromosome, as a homologous pair or as a translocated segment (SCHOENENBERGER *et al.* 2005).

Individual BC1S1-11 was resistant to glufosinate, showing that a wheat transgene can be expressed in an *Ae. cylindrica* background, moreover this is the first generation where it could be present in a homozygous state.

Fertility: BC1 plants can be self fertile and produce seeds. It is therefore not required to have two additional crosses after hybridization to recover partially self-fertile plants as stated by ZEMETRA *et al.* (1998). In fact, SNYDER *et al.* (2000) obtained partially self-fertile BC1s to either parental species and seed set averaged 0.06% and 0.3% in two years; BC1 individuals having *Ae. cylindrica* as the recurrent parent could have as much as 46% of the florets bearing seeds by selfing. Unfortunately, chromosome numbers of these plants are unknown, but could be close to euploid *Ae. cylindrica*. Production of the BC2 generation has been defined as being critical from the standpoint of gene flow, because it could serve as pollen donor to *Ae. cylindrica* and it could propagate even in absence of the progenitor species if self fertility is restored (WANG *et al.* 2001). We show that in some cases, particularly in plants having chromosome numbers close to euploid *Ae. cylindrica*, gene flow may need less backcrossing generations to occur. BC1 plants having 28 or 29 chromosomes were found to be fertile under open pollination conditions in presence of pure *Ae. cylindrica* in a previous study (GUADAGNUOLO *et al.* 2001). It cannot be excluded that a proportion of the offspring of these plants were produced by selfing.

BC1S1 plants having 70-80% of their florets bearing seed may be considered as fully fertile if compared with pure *Ae. cylindrica*.

Fertility rates of the hybrids and germination of BC1 seed are low. Nevertheless, under severe selection pressure as in the case of herbicide resistant GM-wheat cultures invaded by *Ae. cylindrica*, resistances in the wild species due to gene flow, would provide a high selective advantage to introgressed hybrids and successive generations. As genes located on the three genomes of wheat may introgress at least up to BC1S1, the insertion of the transgene on A or B genome instead of D genome is not a sufficient containment measure.

Evidence of past introgression: Researches on introgression between wheat and *Ae. cylindrica* focus exclusively on present mechanisms of introgression, while no literature documents past gene flow. Nevertheless, CALDWELL *et al.* (2004) investigated Single Nucleotide Polymorphisms (SNP) in a coding region of the D genome in *Ae. tauschii*, *Ae. cylindrica*, and wheat. The authors detected 12 different haplotypes, 11 in *Ae. tauschii*, two in *Ae. cylindrica* and two in wheat. They concluded that different haplotypes of the diploid D genome ancestor *Ae. tauschii* contributed to the formation of allotetraploid *Ae. cylindrica*. Surprisingly, haplotype-1, present in all three species, was the most frequent haplotype in wheat (98,5%) and the least frequent in *Ae. cylindrica* (4,3%, of 70 accessions). This pattern is typical from gene flow and suggests that presence of haplotype-1 in both species could be the consequence of introgression rather than of common ancestry. If true, this study would be the first report on past gene flow from wheat to *Ae. cylindrica* and would assess that introgression that was demonstrated in our present study spontaneously occurred in the past.

Acknowledgments

We are grateful to Christof Sautter of the Institute of Plant Sciences, Federal Institute of Technology Zürich, Switzerland, for providing seeds of the wheat varieties Greina and Golin and for his comments on the manuscript; to Eva Stoger of the Institute for Biology of the RWTH Aachen, Germany, for providing seeds of the wheat variety Bobwhite; to Sarah Mamie and Xavier Berney for technical assistance; and to the team of the Botanical Garden of the University and the City of Neuchâtel for taking care of the plants.

This project was funded by the National Centre of Competence in Research (NCCR) Plant Survival, a research program of the Swiss National Science Foundation.

References

ABBOTT, R. J., 1992 Plant invasions, interspecific hybridization and the evolution of new plant taxa. Trends Ecol. Evol. 7: 401-405.

ABRANCHES, R., A. P. SANTOS, E. WEGEL, S. WILLIAMS, A. CASTILHO, *et al.*, 2000 Widely separated multiple transgene integration sites in wheat chromosomes are brought together at interphase. Plant J. 24: 713-723.

ANDERSON, J. A., L. MATTHIESEN, AND J. HEGSTAD, 2004 Resistance to an imidazolinone herbicide is conferred by a gene on chromosome 6DL in the wheat line cv. 9804. Weed Sci. 52: 83-90.

BELEA, A., 1968 Examination of F1 Hybrids of *Aegilops Cylindrica* Host X *Triticum aestivum* L. Acta. Agron. Hung. 17: 151-160.

CADALEN, T., C. BOEUF, S. BERNARD, AND M. BERNARD, 1997 An intervarietal molecular marker map in *Triticum aestivum* L em Thell and comparison with a map from a wide cross. Theor. Appl. Genet. 94: 367-377.

CALDWELL, K. S., J. DVORAK, E. S. LAGUDAH, E. AKHUNOV, M. C. LUO, *et al.*, 2004 Sequence polymorphism in polyploid wheat and their D-genome diploid ancestor. Genetics 167: 941-947.

CHASSOT, P., S. NEMOMISSA, Y. M. YUAN, AND P. KÜPFER, 2001 High paraphyly of *Swertia* L. (Gentianaceae) in the *Gentianella*- lineage as revealed by nuclear and chloroplast DNA sequence variation. Plant Syst. Evol. 229: 1-21.

CLAUSEN, M., R. KRÄUTER, G. SCHACHERMAYR, I. POTRYKUS, AND C. SAUTTER, 2000 Antifungal activity of a virally encoded gene in transgenic wheat. Nat. Biotechnol. 18: 446-449.

DAVID, J. L., E. BENAVENTE, C. BRÈS-PATRY, J.-C. DUSAUTOIR, AND M. ECHAIDE, 2004 Are neopolyploids a likely route for a transgene walk to the wild? The *Aegilops ovata* x *Triticum turgidum durum* case. Biol. J. Linn. Soc. 82: 503-510.

DONALD, W. W., AND A. G. OGG, 1991 Biology and control of jointed goatgrass (*Aegilops cylindrica*), a review. Weed Technol. 5: 3-17.

ELLSTRAND, N. C., AND K. A. SCHIERENBECK, 2000 Hybridization as a stimulus for the evolution of invasiveness in plants? Proc Natl Acad Sci USA. 97: 7043-50.

GILL, B. S., B. FRIEBE, AND T. R. ENDO, 1991 Standard Karyotype and Nomenclature System for Description of Chromosome Bands and Structural-Aberrations in Wheat (*Triticum aestivum*). Genome 34: 830-839.

GUADAGNUOLO, R., D. SAVOVA-BIANCHI, AND F. FELBER, 2001 Gene flow from wheat (*Triticum aestivum* L.) to jointed goatgrass (*Aegilops cylindrica* Host.), as revealed by RAPD and microsatellite markers. Theor. Appl. Genet. 103: 1-8.

HEGDE, S. G., AND J. G. WAINES, 2004 Hybridization and introgression between bread wheat and wild and weedy relatives in North America. Crop Sci. 44: 1145-1155.

KIMBER, G., AND Y. H. ZHAO, 1983 The D genome of the Triticeae. Can. J. Genet. Cytol. 25: 581-589.

KROISS, L. J., P. TEMPALLI, J. L. HANSEN, M. I. VALES, O. RIERA-LIZARAZU, *et al.*, 2004 Marker-assessed retention of wheat chromatin in wheat (*Triticum aestivum*) by jointed goatgrass (*Aegilops cylindrica*) backcross derivatives. Crop Sci. 44: 1429-1433.

LADIZINSKY, G., 1985 Founder effect in crop-plant evolution. Econ. Bot. 39: 191-199.

LIN, Y., 2001 Risk assessment of bar gene transfer from B and D genomes of transformed wheat (*Triticum aestivum*) lines to Jointed goatgrass (*Aegilops cylindrica*). J. Anhui Agric. Uni. 28: 115-118.

LINC, G., B. R. FRIEBE, R. G. KYNAST, M. MOLNAR-LANG, B. KÖSZEGI, *et al.*, 1999 Molecular cytogenetic analysis of *Aegilops cylindrica* Host. Genome 42: 497-503.

MORRISON, L. A., L. CRÉMIEUX, AND C. A. MALLORY-SMITH, 2002a Infestations of jointed goatgrass (*Aegilops cylindrica*) and its hybrids in Oregon wheat fields. Weed Sci. 50: 737-747.

PAILLARD, S., T. SCHNURBUSCH, M. WINZELER, M. MESSMER, P. SOURDILLE, *et al.*, 2003 An integrative genetic linkage map of winter wheat (*Triticum aestivum* L.). Theor. Appl. Genet. 107: 1235-1242.

PETIT, C., F. BRETAGNOLLE, AND F. FELBER, 1999 Evolutionary consequences of diploid-polyploid hybrid zones in wild species. Trends Ecol. Evol. 14: 306-311.

PRIADCENCU, A., C. MICLEA, AND L. MOISESCU, 1967 The local form of the species of *Aegilops cylindrica* Host. and its genetic importance. Rev. Roum. Biol. Ser. Bot. 12: 421-425.

RAJHATHY, T., 1960 Continuous spontaneous crosses between *Aegilops cylindrica* and *Triticum aestivum*. Wheat Inf. Ser. 11: 20.

SAVOVA-BIANCHI, D., 1996 *Evaluation of gene flow between crops and related weeds: risk assessment for releasing transgenic barley (Hordeum vulgare L.) and Alfalfa*

(*Medicago sativa* L.) *in Switzerland.* PhD thesis, University of Neuchâtel, Switzerland.

SCHOENENBERGER, N., F. FELBER, D. SAVOVA-BIANCHI, AND R. GUADAGNUOLO, 2005 Introgression of wheat DNA markers from A, B and D genomes in early generation progeny of *Aegilops cylindrica* Host x *Triticum aestivum* L. hybrids. Theor. Appl. Genet. 111: 1338-1346.

SEEFELDT, S. S., R. ZEMETRA, F. L. YOUNG, AND S. S. JONES, 1998 Production of herbicide resistant jointed goatgrass (*Aegilops cylindrica*) x wheat (*Triticum aestivum*) hybrids in the field by natural hybridization. Weed Sci. 46: 632-634.

SNYDER, J. R., C. A. MALLORY-SMITH, S. BALTER, J. L. HANSEN, AND R. S. ZEMETRA, 2000 Seed production on *Triticum aestivum* by *Aegilops cylindrica* hybrids in the field. Weed Sci. 48: 588-593.

STOGER, E., S. WILLIAMS, P. CHRISTOU, R. E. DOWN, AND J. A. GATEHOUSE, 1999 Expression of the insecticidal lectin from snowdrop (*Galanthus nivalis* agglutinin; GNA) in transgenic wheat plants: effects on predation by the grain aphid *Sitobion avenae*. Mol. Breeding 5: 65-73.

TAKUMI, S., AND T. SHIMADA, 1996 Production of transgenic wheat through particle bombardment of scutellar tissues: Frequency is influenced by culture duration. J. Plant Physiol. 149: 418-423.

VAN SLAGEREN, M. W., 1994 *Wild wheats: a monograph of Aegilops L. and Amblyopyrum (Jaub. & Spach) Eig (Poaceae)*. Wageningen Agricultural University Press, Wageningen; ICARDA, Aleppo.

WAINES, J. G. AND D. BARNHART, 1992 Biosystematic Research in *Aegilops* and *Triticum*. Hereditas 116: 207-212.

WANG, Z., R. S. ZEMETRA, J. HANSEN, AND C. A. MALLORY-SMITH, 2001 The fertility of wheat x jointed goatgrass hybrid and its backcross progenies. Weed Sci. 49: 340-345.

WANG, Z. N., A. HANG, J. HANSEN, C. BURTON, C. A. MALLORY-SMITH, *et al.*, 2000 Visualization of A- and B-genome chromosomes in wheat (*Triticum aestivum* L.) x jointed goatgrass (*Aegilops cylindrica* Host) backcross progenies. Genome 43: 1038-44.

WANG, Z. N., R. S. ZEMETRA, J. HANSEN, A. HANG, C. A. MALLORY-SMITH, *et al.*, 2002 Determination of the paternity of wheat (*Triticum aestivum* L) x jointed goatgrass (*Aegilops cylindrica* host) BC1 plants by using genomic in situ hybridization (GISH) technique. Crop Sci. 42: 939-943.

WHITE, J., S.-Y. P. CHANG, M. J. BIBB, AND M. J. BIBB, 1990 A cassette containing the bar gene of *Streptomyces hygroscopicus*: a selectable marker for plant transformation. Nucleic Acids Res. 18: 1062.

ZEMETRA, R. S., J. HANSEN, AND C. A. MALLORY-SMITH, 1998 Potential for gene transfer between wheat (*Triticum aestivum*) and jointed goatgrass (*Aegilops cylindrica*). Weed Sci. 46: 313-317.

CHAPTER 4

RAPDs reveal low genetic variability in natural *Aegilops cylindrica* Host populations in Europe and Northern America

Abstract

The close wheat relative, *Aegilops cylindrica* Host (Poaceae) is a potential source of genetic variation for wheat improvement and an important agricultural weed, particularly in wheat fields. Twenty-three *Ae. cylindrica* populations, most of them from adventive locations in Switzerland, Italy and the USA were investigated using random amplified polymorphic DNAs (RAPDs). A total of 380 *Ae. cylindrica* individuals were analysed using 9 decamer RAPD primers and yielded 70 clear and repeatable RAPD markers, of which 34.3% were polymorphic. Genetic diversity of *Ae. cylindrica* was low. As it is typical for inbreeding species, most genetic variance resided among populations (ΦST: 0.91). Genetic pattern of the populations did not necessarily correlate with geographic distribution; a Romanian population was the most distinct one, whereas Northern American populations mainly clustered together with populations from the Italian Aosta valley. Swiss populations were mostly distinct compared to Italian ones. Nevertheless, a very recent population from a Swiss vineyard that underwent a drastic increase in population size these last few years, was genetically identical to a weedy population from Southern California. *Ae. cylindrica* may have migrated several times across the Atlantic in both directions, together with humans or goods.

Introduction

The genus *Aegilops* L. (Poaceae) is a member of the Triticeae tribe, which is one of the plant groups which has been most intensively investigated (Miller 1987). *Aegilops* comprises 22 wild annual species and is very closely related to cultivated wheats (*Triticum* L.) (Van Slageren 1994), due to its involvement in the evolution of polyploid wheats. *Aegilops cylindrica* Host is a Mediterranean-Western Asiatic species, common throughout its natural distribution range, its western boundaries are along the Black Sea coast and along the Danube river basin up to Hungary (Van Slageren 1994; Zaharieva et al. 2004). Adventive locations of *Ae. cylindrica* are found in ruderal habitats in Western and Northern Europe like railway areas, harbours, roadsides or vineyards (Guadagnuolo et al. 2001a; Martini and Pericin 2003; Schoenenberger and Giorgetti-Franscini 2004). It was found in wheat fields in Hungary (Rajhathy 1960), from where the hybrid between *Ae. cylindrica* and wheat was first described under the name x *Aegilotriticum sancti-andreae* (Degen) Soó (Soó 1951). In Europe and particularly in Switzerland, *Ae. cylindrica* is considered to be a rare species and is included in Switzerland's Red List of Threatened Taxa and classed in the IUCN category VU (vulnerable)(Moser et al. 2002).

Ae. cylindrica was repeatedly introduced to Northern America as a seed contaminant at the end of the nineteenth century and soon became a serious agricultural weed, particularly in wheat fields. On a historical base, it is assumed that the introduced seed originated from the Eastern Mediterranean (reviewed in Donald and Ogg 1991). In fact, among all *Aegilops* species, *Ae. cylindrica* is the one displaying the most weedy behaviour (Van Slageren 1994). In Northern America, serious economic loss in wheat cultivation is caused by competition in winter wheat fields, with the consequence of reduced crop yields and grain quality loss because of seed contamination. Moreover, because of the genetic relatedness between wheat and *Ae. cylindrica*, no selective herbicides to control the weed can be used (Donald and Ogg 1991).

Ae. cylindrica is an allotetraploid (2n=4x=28, CCDD genome) and shares the D genome ancestor *Aegilops tauschii* Coss. with bread wheat (*Triticum aestivum* L., 2n=6x=42, AABBDD genome) (Kimber and Zhao 1983). Due to the close phylogenetic relationship between the two species, an exchange of genetic material by natural hybridisation and subsequent introgression is possible. This fact is particularly relevant for risk assessment of herbicide resistant wheat, either transgenic, mutation or conventionally bred, in the case of gene flow from the crop to *Ae. cylindrica* (Guadagnuolo et al. 2001a; Schoenenberger et al. 2005; Seefeldt et al. 1998; Wang et al. 2000). On the other hand, the fact that *Ae. cylindrica* is part of the wheat secondary gene pool (Harlan and de Vet 1971), makes it an interesting source of genes for wheat improvement. Traits of *Ae. cylindrica* which may be interesting for breeding programs include abiotic stress resistances like the species' elevated frost tolerance (Zaharieva et al. 2003) and salt tolerance (Farooq and Farooq 2001), and its resistance to fungal pathogens like speckled snow mold (Iriki et al. 2001).

Due to the species' evident economic interests, several studies on the genetic diversity in *Ae. cylindrica* have been implemented by the use of RAPDs (Goryunova et al. 2004; Okuno et al. 1998; Pester et al. 2003), AFLPs (Pester et al. 2003), isozymes (Hegde et al. 2002; Watanabe et al. 1994), chromosome C-banding (Badaeva et al. 2002; Linc et al. 1999) and microsatellites (Gandhi et al. 2005). Interestingly, all these studies draw a similar conclusion: genetic variability in *Ae. cylindrica* is very low. Moreover, genetic variability is lower when compared to its diploid ancestors *Ae tauschii* Coss. (DD genome) and *Ae caudata* L. (CC genome) or to other tetraploid *Aegilops* species. The work presented here is the first aimed at detecting the genetic structure and differentiation of natural *Ae. cylindrica* populations on the base of a large sampling both in Europe and in the USA. We aim to achieve a better understanding of the genetic variation and get insights on the migration routes taken by this extremely interesting species.

Materials and methods

Plant material: In 2002, we searched for natural populations of *Aegilops cylindrica* Host in Switzerland, Italy (Aosta valley) and in the USA. Several populations were found by referring to the literature and by asking Swiss botanists (Table 1). Population CH 4 was discovered by Beatrice Moor, St. Galler-Ring 192, 4054 Basel, Switzerland; CH 6 by Ralph Imstepf, Stipa - Beratung für Natur und Landschaft, 3953 Leuk-Stadt, Switzerland; and CH 7 by Yann Clavien, Laboratory for Soil and Vegetation Science of the University of Neuchâtel, Switzerland. Populations USA 5, 6 and 7 were discovered and seeds collected by Dieter and Anke Burger, NW 1520 Kenny Drive, Pullman 99163v, WA, USA. Seeds of individual plants were sampled and up to 25 seeds per population originating from distinct mother plants were sown in pots at the botanical garden of Neuchâtel, Switzerland. If the population was smaller, one descendant of each plant growing in nature was analysed. Furthermore, seeds from some of the populations were obtained from seeds banks, accession numbers are given in Table 1. A total of 380 *Ae. cylindrica* individuals were analysed.

Table1. sampling locations and habitats of *Ae. cylindrica* populations and amount of individuals analysed

Population name	Geographic origin	Habitat	Citation or accession number	sample size
CH1	Brig 7° 59' 3" E; 46° 19' 13" N	railway	Becherer 1956	12
CH2	Sierre 7° 28' 55" E; 46° 16' 24" N	roadside	Guadagnuolo et al. 2001a	20
CH3	Saillon 7° 11' E; 46° 10' 14" N	roadside	Guadagnuolo et al. 2001a	9
CH4	Basel 7° 36' 5" E; 47° 34' 45" N	railway	Brodtbeck et al. 1998	21
CH5	Riazzino-Cugnasco 8° 54' 1" E; 46° 10' 27" N	railway	Schoenenberger and Giorgetti 2004	18
CH6	Sierre 7° 32' 54" E; 46° 17' 52" N	roadside		20
CH7	Miège 7° 32' 16" E; 46° 18' 58" N	vineyard		4
CH8	Eyholz 7° 55' 39" E; 46° 17' 55" N	roadside		3
Ao1 (Italy)	Sarre 7° 14' 48" E; 45° 43' 8" N	roadside	Pignatti 1982	19

(continued)

Table 1. *Continued*

Population name	Geographic origin	Habitat	Citation or accession number	sample size
Ao2 (Italy)	Monte Torretta 7° 14' 23" E; 45° 43' 3" N	roadside	Pignatti 1982	19
Ao3 (Italy)	Croix 7° 13' 55" E; 45° 43' 2" N	vineyard	Pignatti 1982	20
Ao4 (Italy)	Vereytaz 7° 13' 2" E; 45° 42' 48" N	roadside	Pignatti 1982	20
Ao5 (Italy)	Sarre Gare 7° 15' 30" E; 45° 43' 5" N	railway	Pignatti 1982	20
EU1 (France)	Tallard 6° 03' E; 44° 28' N		91-33 Museum d'Histoire Naturelle Paris	13
EU2 (Romania)	Vama Veche 28° 34 E; 43° 45 N		2230 Botanical Garden Cluj-Napoca	10
USA1 (CA)	Rim of the world Hgw. 117° 17' 42" W; 34° 13' 33"	roadside		19
USA2 (CA)	Crestline 117° 18' 50" W; 34° 14' 52" N	roadside	Hegde et al. 2002	20
USA3a (ID)	Newdale 111° 35 W; 43° 45' N	farm, grain bin	PI 502242 NSGC USDA	13
USA3b (ID)	Newdale 111° 35 W; 43° 45' N	farm, parking area	PI 502243 NSGC USDA	7
USA4 (WA)	Treasureton 111° 53' W; 42° 16' N	farm	PI 506231 NSGC USDA	13
USA5 (WA)	Albion 117° 16' 33" W; 46° 46' 22" N	wheat field		20
USA6 (WA)	Albion 117° 15' 46" W; 46° 47 13" N	road next to wheat field		40
USA7 (WA)	Pullman 117° 11' 6" W; 46° 43' 55" N	railway		20

DNA extraction and RAPD: Genomic DNA was extracted from frozen leaf tissue (-80°C) of individual *Ae. cylindrica* plants following a simple and cheap SDS-Na-acetate protocol (Savova-Bianchi 1996). DNA was resuspended in a TE (pH 8) buffer, concentration adjusted to 20-40 ng/µl and stored at -20°C. A preliminary screen of 80 RAPD primers was carried out on 13 individuals originating from distinct populations of all geographical areas, in order to define primers showing highest polymorphism between populations and clearest reading of the bands on agarose gels. Decamer primers of the series OPB, OPC, OPF, OPG, OPP and OPT (Operon Technologies, Alameda, CA, USA) were screened for that purpose. RAPD reactions were carried out in a volume of 25µl containing 1x PCR buffer, 1.5 mM $MgCl_2$, 0.4x Q-solution (Qiagen), 0.2 mM dNTP, 0.2 µM primer, 0,03 U/µl

Taq polymerase and 1 ng/µl template DNA. Amplification reactions were carried out in a Biometra T3 thermocycler. Cycling parameters were: initial denaturation at 94°C for 10 min, followed by 40 cycles of 1 min at 93°C, 1 min at 41-45°C depending on the primer used, and 1 min at 72°C. Final extension was 72°C for 10 min. Primers chosen for the analysis of the whole sample set were OPB8 (44°C annealing temperature), OPB10 (45°C), OPC2 (43°C), OPC4 (41°C), OPF 14 (42°C), OPG5 (41°C), OPG16 (41°C), OPP8 (45°C) and OPP9 (45°C) (Table 2). In order to test reproducibility of the results, RAPDs were repeated twice in a subset of 50-100 individuals for each primer. Amplification products were mixed with 1/5 loading buffer and loaded onto 1,4% (w/v) agarose gels stained with ethidium bromide. Electrophoresis was carried out at 100 V. RAPD fragments were photographed under UV light.

Table 2. RAPD primers employed and number of scored fragments

Primers	Sequence (5'→ 3')	Number of fragments		
		Polymorphic	Monomorphic	Total
OPB8	GTCCACACGG	5	6	11
OPB10	CTGCTGGGAC	1	7	8
OPC2	GTGAGGCGTC	2	7	9
OPC4	CCGCATCTAC	3	3	6
OPF14	TGCTGCAGGT	4	3	7
OPG5	CTGAGACGGA	1	5	6
OPG16	AGCGTCCTCC	5	3	8
OPP8	ACATCGCCCA	1	6	7
OPP9	GTGGTCCGCA	2	6	8

Data analysis: Repeatable and clearly distinguishable RAPD bands were scored for each *Ae. cylindrica* individual as presence or absence. The CLUSTER package (http://www2.biology.ualberta.ca/jbrzusto/cluster.php) was used to calculate an euclidean distance matrix, on the base of which an UPGMA (Unweighted Pair Group Method with Arithmetic Mean) clustered dendrogram was generated, and graphically visualized with TREEVIEW (Page 1996). As genetic polymorphism between *Ae. cylindrica* individuals was very low, we considered that a common absence of a RAPD fragment between two individuals represents an element of similarity, in the same way as a common presence of a fragment. We used then an euclidean distance and not a jaccard's coefficient as one would use if different species were compared (Guadagnuolo et al. 2001b). As the dendrogram containing all 380 *Ae. cylindrica* individuals was simply too big to be represented in a graphically readable way, individuals of the same population, showing no polymorphism, were pooled together. All individuals which were genetically distinct or belonging to different populations were kept separately for the analysis. By this way, a dendrogram readable on a single page, maintaining the same clades as the original dendrogram, was obtained. The same pooled matrix was used to perform principal coordinates analysis (PCoA). Graphic output was generated with StatView 4.51 (Abacus concepts, Inc. Berkeley, CA, USA). All other statistical analyses, including populations genetics, were carried out with the whole sample set. Statistical significance of the clustered RAPD generated groups and their discrimination between populations and geographic distributions, was performed trough Mantel tests (999 permutations), using the R4 (Beta version) package (P. Casgrain and P. Legendre, Département des Sciences Biologiques, Université de Montreal, Québec, Canada). Genetic distance matrix was confronted to a hypothesis testing matrix, where minimum distance (0) was assigned to individuals of groups of populations from the same geographical origin (France, Italy, Switzerland, Romania USA) and maximum distance (1) assigned to individuals of different geographical origins. Genetic diversity data analyses were performed using ARLEQUIN version 2.0 (S. Schneider, D. Roessli and L. Excoffier, Genetics and Biometry Laboratory, Department of Anthropology and Ecology, University of Geneva, Switzerland). Analysis of molecular variance (AMOVA) package was used to partition the variation within and among populations and regions.

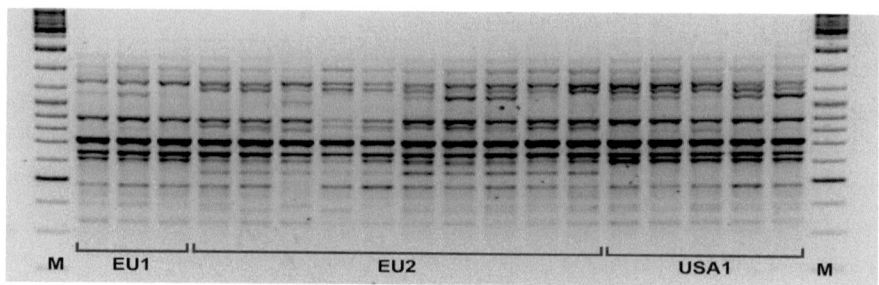

Figure 1. Example of an RAPD amplification with primer OPB8 showing polymorphisms among *Ae. cylindrica* individuals from populations in France (EU1) Romania (EU2) and California (USA1). M: 100 bp ladder.

Results

A total of 380 *Ae. cylindrica* individuals originating from 23 populations were analysed using 9 decamer RAPD primers (Fig 1). Seventy clear and repeatable RAPD fragments were scored, of which 24 (34.3%) were polymorphic (Table 2); 38 fragments were present in all *Ae. cylindrica* individuals. 8 were present or absent in less than 1% of the individuals out of 380 and thus not very informative. All other fragments showed polymorphism among larger groups of individuals or populations. The number of fragments generated per primer varied between 6 and 11, among these 14.3% to 62.5% were polymorphic (Table 2). An UPGMA dendrogram clustered 72 *Ae. cylindrica* individuals representing all detected genetic diversity in distinct groups which do not necessarily reflect a geographical distribution (Fig 2). In fact there is a significant correlation of only 0.31 between the genetic distance matrix and a matrix were maximum differentiation between geographical areas is postulated. Similarly, populations were not well differentiated between each other (Table 3).

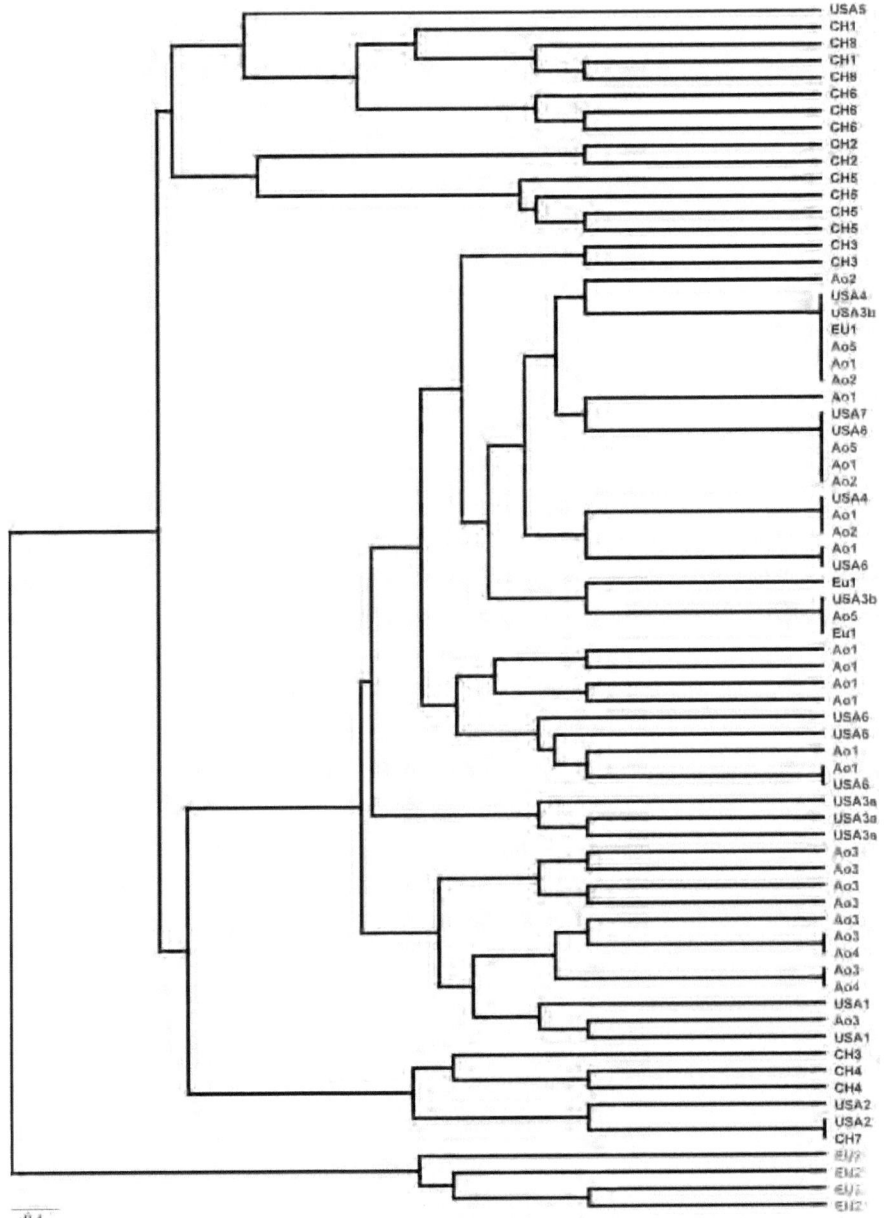

Figure 2. UPGMA dendrogram based upon the RAPD similarity matrix (using an Euclidean distance)

Table 3. r-statistics calculated from the Mantel permutation test. Correlations were highly significant (*: P < 0.01)

Comparison	Correlation
Genetic euclidean dissimilarity versus maximum dissimilarity between populations	0.45*
Genetic euclidean dissimilarity versus maximum dissimilarity between geographic areas (Switzerland, Italy, France, Romania, USA)	0.31*

The population from Romania (EU2) is the most distinct one and it clusters basally to all other populations. A second clade contains several Swiss genotypes (CH) and a Northern American (USA5) genotype. Interestingly, RAPDs do not reveal genetic differentiation among several American populations and populations from Switzerland (USA2 and CH7) or from the Italian Aosta valley (Ao1, 2, 5 and USA3b and 4). However, there is a quite clear differentiation between Swiss and Italian populations as it is visible in principal coordinates analysis where the data is plotted onto a bidimensional scattergram (Fig 3). The northern American populations are scattered throughout the Italian group, and to a lesser extent through the Swiss group. Here again, the Romanian population is the most distinct one, whereas the French population is close to the Italian ones. However, the two axes of the scattergram represented only about 22% of total variance, whereas the first three axes explain about 30% of total variance. As revealed by the analysis of molecular variance (AMOVA, Table 4), most of the variation generated with the RAPD markers resides among populations. Globally, about 69% of the variation was within populations of the same regions, 22% was among regions and only 9% was within populations. Within population variation is always below 10% whether single geographic areas are taken into account or all of them. By comparing only Swiss and Italian populations (CH & Ao) 27% of the variation resides among regions, whereas if American population data is included, the value is as low as 3%.

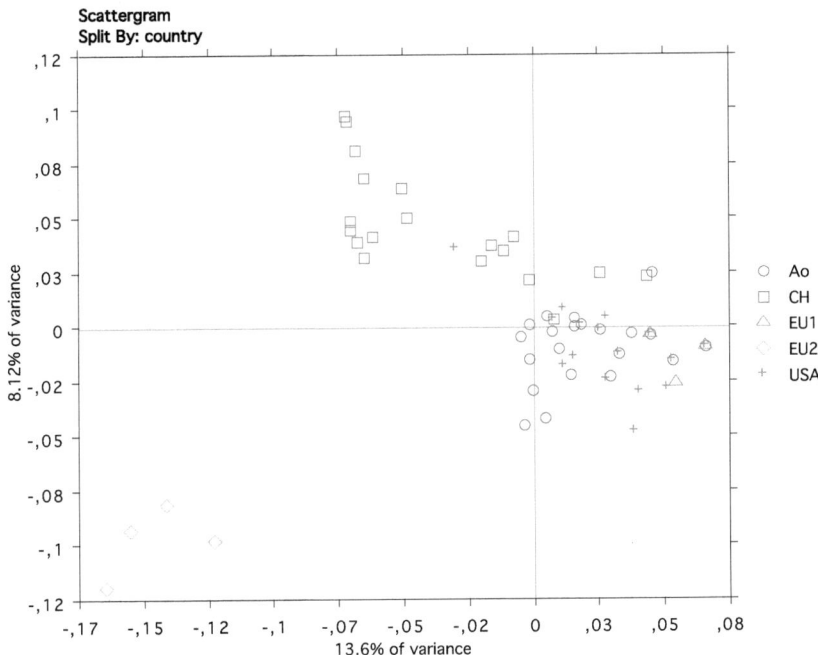

Figure 3. Principal Coordinates Analysis (PCoA) of the RAPD based Euclidean distance matrix

Table 4. AMOVA analyses based on RAPD-distances. Phi (φ) statistics defined by three fixation indexes: ΦCT is the proportion of differentiation between regions, φSC is the differentiation among populations within a region and ΦST the global differentiation of populations. Significance tests (1023 permutations) ***$P<0.0001$, **$P<0.005$, *NS* non significant, d.f. degrees of freedom

Source of variation	d.f.	Sum of Squares	Variance components	Percentage of variation	Fixation index
CH					
among populations	7	301.17	3.33	94.06	φST: 0.94***
within populations	99	20.79	0.21	5.94	
CH & Ao					
among regions	1	137.2	0.94	26.81	φSC: 0.88***
among populations					φST: 0.91***
within groups	11	381.27	2.24	64.28	φCT: 0.27**
within populations	192	59.71	0.31	8.91	
USA					
among Populations	7	234.54	1.76	91.04	φST: 0.91***
within Populations	144	24.91	0.17	8.96	
CH + Ao & USA					
among regions	1	57.6	0.076	2.84	φSC: 0.90***
among Populations					φST: 0.91***
within groups	19	753.01	2.34	87.72	φCT: 0.03NS
within Populations	336	84.61	0.25	9.43	
All					
among regions	4	294.79	0.65	22.05	φSC: 0.88***
among populations					φST: 0.91***
within regions	18	615.81	2.03	68.93	φCT: 0.22**
within populations	357	94.71	0.27	9.01	

Discussion

As it is typical for annual, selfing and early successional taxa, most genetic variability resides among populations (Hamrick 1979; Nybom 2004) and is confirmed by the genetic structure of *Ae. cylindrica* populations (Table 4). Detected genetic diversity was low in *Ae. cylindrica* populations in adventive locations in the USA and Western Europe, as well as in a natural location on the Romanian Black Sea coast. This is consistent with isozyme comparisons carried on at least 30 individuals per population in two natural sites, one from California and the other from Lebanon. No genetic differentiation was detected within nor among the populations (Hegde et al. 2002). Similarly, by comparing alpha amylase zymograms of 453 *Ae. cylindrica* accessions covering the whole distribution range of the species, 441 were completely monomorphic, the remaining accessions were variable, suggesting that the Eastern Mediterranean and Iran were the centres of origin of the species (Watanabe et al. 1994). However, by screening a relatively high number of RAPD primers, we could identify nine primers that amplified 24 fragments (34%) mainly displaying between-population polymorphisms. The yield of polymorphic bands was thus higher than in another study where RAPDs were used to differentiate *Ae. cylindrica* accessions (Pester et al. 2003). The authors identified 6.7% polymorphic bands using 30 RAPD primers by analysing Northern American and Eurasian *Ae. cylindrica* accessions, and hypothesized multiple introductions into the USA or genetic divergence since introduction. Interestingly, by the use of RAPDs for comparing single individual *Ae. cylindrica* accessions from North Caucasia and Central Asia, it was established that genetic diversity in the latter region (89% polymorphic bands) was much higher than in the former (30% polymorphic bands) (Okuno et al. 1998), both regions being part of the species' natural distribution range (Van Slageren 1994). Moreover, Central Asian accessions of *Ae. cylindrica* were intermingled in the same cluster as *Ae. tauschii* from the same region, whereas in Northern Caucasia the species were clearly separated into distinct clusters (Okuno et al. 1998). The authors supposed that North Caucasia may be the edge of the distribution of *Ae. cylindrica* and thus less diverse. An alternate explanation may be that hybridization between the two species is taking place in Central Asia and that the *Ae. cylindrica* gene pool is enriched by genetic variability coming from its diploid D genome ancestor. Nevertheless, it is not probable that Central Asia is the centre of origin of *Ae. cylindrica* as the C genome diploid ancestor *Ae. caudata* L. is absent from that region (Van Slageren 1994).

American *Ae. cylindrica* populations are not less diverse than European ones. An explanation for this genetic pattern may be population differentiation in the USA since introduction. A global loss of genetic diversity due to founder effect could have deep effects on the genetic structure of the populations, on both continents. On the other hand, multiple introductions of invasive *Ae. cylindrica* in USA as postulated by Donald and Ogg (1991) may be a better interpretation, particularly if historical factors as extensive human migrations from Europe to the USA in the 19th and beginning of the 20th century are taken into account. This seems particularly true in consideration of the similarity between populations from the USA and the Italian Aosta Valley, which was for long time an area from which people used to emigrate to Northern America and possibly came back. However, it may be difficult to tell whether the Aosta valley was one of the origins of synanthropic *Ae. cylindrica* introduced in the USA or if seeds from Northern America were brought back to that region. However, most Italian populations are more closely related to American ones than to Swiss ones (Fig 2 & 3), which is surprising considered the fact that geographic distance between Wallis and the Aosta valley is only about 50 kilometres. Interestingly, individuals from the Californian population USA2, corresponding to population 11 in Hedge et al. (Hegde et al. 2002) were genetically identical to a recently introduced population in a vineyard in Wallis (Population CH7), which had only 4 individuals in 2003 but underwent a big increase in individual number and population surface by 2005 (Yann Clavien, pers. comm.). Moreover, several new populations of *Ae. cylindrica* were discovered in Wallis by Swiss botanists during the present investigation and after it had started. Compared to a previous campaign (Guadagnuolo et al. 2001a), these new populations are indicating a recent spread of the weedy species in Switzerland. On the other hand, the Swiss populations of Brig (CH1) which is known since 1938 (Becherer 1956), and Saillon (CH3), which is also known since a long time, were always stable in size and area occupied.

The evidence that *Ae. cylindrica* has low genetic variation and is a successful invader is similar to other species occurring in ruderal habitats which have been introduced repeatedly to Northern America. As an example, *Bromus tectorum* L. also shows reduced genetic variability and benefits of an extremely high invasion success (Novak et al. 1991; Novak et al. 1993). Similarly, a lower genetic variation was observed for weedy *Secale cereale* L. in Southern California (Hegde et al. 2002).

In investigations that address the question of intraspecific diversity and draw phylogeographic conclusions it is important to use material directly collected in nature as it was done by Hedge et al. (Hegde et al. 2002), particularly in the case of annual species like *Ae. cylindrica* as it was shown that reproduction cycles in germplasm collections can seriously alter allele frequencies compared to original material particularly in open pollinated species (Chebotar et al. 2003).

Acknowledgments

We are grateful to Beatrice Moor, Yann Clavien and Ralph Imstepf for indicating Swiss *Ae. cylindrica* populations and to Dieter and Anke Burger for collecting seeds in Washington USA. Sarah Mamie is greatly acknowledged for the excellent laboratory work she did, while Nils Arrigo improved the quality of the manuscript with his useful comments. We thank also Philippe Küpfer for his support. This project was funded by the National Centre of Competence in Research (NCCR) Plant Survival, a research programme of the Swiss National Science Foundation.

References

Badaeva, E. D., A. V. Amosova, O. V. Muravenko, T. E. Samatadze, N. N. Chikida, A. V. Zelenin, B. Friebe, and B. S. Gill. 2002. Genome differentiation in Aegilops. 3. Evolution of the D-genome cluster. Plant Systematics and Evolution. 231:163-190.

Becherer, A. 1956. Florae Vallesiacae Supplementum. Fretz, Zürich, Memoires de la Société Suisse des Sciences Naturelles. 556p.

Brodtbeck, T., M. Zemp, M. Frei, U. Kienzle, and D. Knecht. 1998. Flora von Basel und Umgebung 1980-1996 Teil II. Liestal.

Chebotar, S., M. S. Roder, V. Korzun, B. Saal, W. E. Weber, and A. Borner. 2003. Molecular studies on genetic integrity of open-pollinating species rye (Secale cereale L.) after long-term genebank maintenance. Theoretical and Applied Genetics. 107:1469-1476.

Donald, W. W. and A. G. Ogg. 1991. Biology and control of jointed goatgrass (*Aegilops cylindrica*), a review. Weed Technology. 5:3-17.

Farooq, S. and E. A. Farooq. 2001. Co-existence of salt and drought tolerance in Triticeae. Hereditas. 135:205-210.

Gandhi, H. T., M. I. Vales, C. J. W. Watson, C. A. Mallory-Smith, N. Mori, M. Rehman, R. S. Zemetra, and O. Riera-Lizarazu. 2005. Chloroplast and nuclear microsatellite analysis of Aegilops cylindrica. Theoretical and Applied Genetics. 111:561-572.

Goryunova, S. V., E. Z. Kochieva, N. N. Chikida, and V. A. Pukhalskyi. 2004. Phylogenetic relationships and intraspecific variation of D-genome Aegilops L. as revealed by RAPD analysis. Russian Journal of Genetics. 40:515-523.

Guadagnuolo, R., D. S. Bianchi, and F. Felber. 2001b. Specific genetic markers for wheat, spelt, and four wild relatives: comparison of isozymes, RAPDs, and wheat microsatellites. Genome. 44:610-21.

Guadagnuolo, R., D. Savova-Bianchi, and F. Felber. 2001a. Gene flow from wheat (*Triticum aestivum* L.) to jointed goatgrass (*Aegilops cylindrica* Host.), as revealed by RAPD and microsatellite markers. Theoretical and Applied Genetics. 103:1-8.

Hamrick, J. L. 1979. Relation between life history characteristics and electrophoretically detectable genetic variation in plants. Ann. Rev Ecol. Syst. 10:173-200.

Harlan, J. R. and J. M. J. de Vet. 1971. Toward a rational classification of cultivated plants. Taxon. 20:509-517.

Hegde, S. G., J. Valkoun, and J. G. Waines. 2002. Genetic diversity in wild and weedy Aegilops, Amblyopyrum, and Secale species - A preliminary survey. Crop Science. 42:608-614.

Iriki, N., A. Kawakami, K. Takata, T. Kuwabara, and T. Ban. 2001. Screening relatives of wheat for snow mold resistance and freeing tolerance. Euphytica. 122:335-341.

Kimber, G. and Y. H. Zhao. 1983. the D genome of the Triticeae. Can. J. Genet. Cytol. 25:581-589.

Linc, G., B. R. Friebe, R. G. Kynast, M. Molnar-Lang, B. Köszegi, J. Sutka, and B. S. Gill. 1999. Molecular cytogenetic analysis of *Aegilops cylindrica* Host. Genome. 42:497-503.

Martini, F. and C. Pericin. 2003. Die Flora des Punto Franco vecchio im Areal des alten Hafens von Triest (NE Italien). Bauhinia. 17:

Miller, T. E. (1987). Systematics and evolution. Wheat Breeding. Its scientific bases. F. G. H. Lupton. London, New York, Chapman and Hall: 1-30.

Moser, D., A. Gygax, B. Bäumler, N. Wyler, and R. Palese. 2002. Liste Rouge des fougères et plantes à fleur menacées de Suisse, Office Fédéral de l'Environnement des Forêts et du Paysage, Bern; Centre du Réseau Suisse de Florstique, Chambésy; Conservatoire et Jardin Botaniques de la Ville de Genève.

Novak, S. J., R. N. Mack, and D. E. Soltis. 1991. Genetic-Variation in Bromus-Tectorum (Poaceae) - Population Differentiation in Its North-American Range. American Journal of Botany. 78:1150-1161.

Novak, S. J., R. N. Mack, and P. S. Soltis. 1993. Genetic-Variation in Bromus-Tectorum (Poaceae) - Introduction Dynamics in North-America. Canadian Journal of Botany-Revue Canadienne De Botanique. 71:1441-1448.

Nybom, H. 2004. Comparison of different nuclear DNA markers for estimating intraspecific genetic diversity in plants. Molecular Ecology. 13:1143-1155.

Okuno, K., K. Ebana, B. Noov, and H. Yoshida. 1998. Genetic diversity of Central Asian and north Caucasian Aegilops species as revealed by RAPD markers. Genetic Resources and Crop Evolution. 45:389-394.

Page, R. D. M. 1996. TreeView: An application to display phylogenetic trees on personal computers. Computer Applications in the Biosciences. 12:357-358.

Pester, T. A., S. M. Ward, A. L. Fenwick, P. Westra, and S. J. Nissen. 2003. Genetic diversity of jointed goatgrass (Aegilops cylindrica) determined with RAPD and AFLP markers. Weed Science. 51:287-293.

Pignatti, S. 1982. Flora d'Italia. Volume terzo. Edagricole, Bologna.

Rajhathy, T. 1960. Continuous spontaneous crosses between *Aegilops cylindrica* and *Triticum aestivum*. Wheat Information Service. 11:20.

Savova-Bianchi, D. 1996. Evaluation of gene flow between crops and related weeds: risk assessment for releasing transgenic barley (*Hordeum vulgare* L.) and Alfalfa (*Medicago sativa* L.) in Switzerland. PhD thesis, University of Neuchâtel, Switzerland.

Schoenenberger, N., F. Felber, D. Savova-Bianchi, and R. Guadagnuolo. 2005. Introgression of wheat DNA markers from A, B and D genomes in early generation progeny of *Aegilops cylindrica* Host X *Triticum aestivum* L. hybrids. Theoretical and Applied Genetics. DOI: 10.1007/s00122-005-0063-7:1-9.

Schoenenberger, N. and P. Giorgetti. 2004. Note floristiche ticinesi: la flora della rete ferroviaria con particolare attenzione alle specie avventizie. Parte II. Boll. Soc. Tic. Sci. Nat. 92:97-108.

Seefeldt, S. S., R. Zemetra, F. L. Young, and S. S. Jones. 1998. Production of herbicide resistant jointed goatgrass (*Aegilops cylindrica*) x wheat (*Triticum aestivum*) hybrids in the field by natural hybridization. Weed Science. 46:632-634.

Slageren, M. W. van 1994. Wild wheats: a monograph of *Aegilops* L. and *Amblyopyrum* (Jaub. & Spach) Eig (Poaceae), Wageningen Agricultural University Press, Wageningen; ICARDA, Aleppo. 512p.

Soó, R. (1951). A Magyar növényvilág kézikönyve (Handbook of the Hungarian flora). Volume II: 939.

Wang, Z. N., A. Hang, J. Hansen, C. Burton, C. A. Mallory-Smith, and R. S. Zemetra. 2000. Visualization of A- and B-genome chromosomes in wheat (Triticum aestivum

L.) x jointed goatgrass (Aegilops cylindrica Host) backcross progenies. Genome. 43:1038-44.

Watanabe, N., K. Mastui, and Y. Furuta (1994). Uniformity of the alpha-Amylase isozymes of *Aegilops cylindrica* Host introduced into North America: comparison with ancestral Eurasian accessions. Proceedings of the 2nd international Triticeae symposium, Logan, Utah, Utah State University. 20-24.

Zaharieva, M., A. Dimov, P. Stankova, J. David, and P. Monneveux. 2003. Morphological diversity and potential interest for wheat improvement of three Aegilops L. species from Bulgaria. Genetic Resources and Crop Evolution. 50:507-517.

Zaharieva, M., J. M. Prosperi, and P. Monneveux. 2004. Ecological distribution and species diversity of Aegilops L. genus in Bulgaria. Biodiversity and Conservation. 13:2319-2337.

CHAPTER 5

Gene flow from Wheat (*Triticum aestivum* L.) to *Aegilops geniculata* Roth.

Abstract

Aegilops geniculata Roth is one of the most widespread species of the genus Aegilops L. The species is closely related to wheat, hybrids between the two species are found where they grow in sympatry. In order to quantify hybridisation and the potential of gene flow from wheat to Ae. geniculata, we carried out field trials with Ae. geniculata plants, originating from four distinct populations and growing in agronomical winter wheat fields. We detected eight hybrids among the 850 surviving offspring of the wild species, and the hybridisation rate varied between 0 and 6.06% on a per field and population basis. Two hybrids grown under open pollination conditions produced two BC1 seeds and average fertility of the hybrids was 2.2%. The BC1 plants had 56 and 55 chromosomes and carried wheat-specific SCARs (sequence characterised amplified regions) located on chromosomes 6A and 6D. Although the BC1 plants did not produce seed due to poor Ae. geniculata pollen availability, the results show that gene flow from wheat to Ae. geniculata may occur naturally. The results are relevant for risk assessment of cultivation of genetically engineered wheat.

Introduction

Allotetraploid Aegilops geniculata Roth ($2n=4x=28$, UUMM genome formula) is native to the Mediterranean basin and Western Asia. It is common in Southern Europe and Northern Africa, but rarer in the Eastern part of its distribution, in Libya, Egypt and Turkey (Van Slageren 1994). Adventive locations of the species may be found in northern Europe; Ae. geniculata was introduced to the Canary islands and to a restricted area in California, USA (Van Slageren 1994).

In the debate over the release of genetically engineered organisms, gene flow and its consequences represents one of the main concerns about growing genetically modified crops (Wolfenbarger and Phifer 2000). Most crop species have the ability to hybridise with their wild relatives where they grow in sympatry; the main consequences of gene flow from a cultivated plant to a wild relative, whether this plant is transgenic or not, are the possibilities of increased weediness or enhanced risk of extinction of the wild (Ellstrand 2003). Research on gene flow from wheat to a wild relative has been conducted extensively in the case of jointed goatgrass (Aegilops cylindrica Host), which is a tenacious weed in Northern American wheat fields. Hybridisation frequencies and dynamics between the species, chromosome numbers, fertilities, as well as the fate of introgressed wheat DNA have been investigated in the perspective of risk assessment for the release of genetically modified wheat (Guadagnuolo et al. 2001a; Morrison et al. 2002a; Morrison et al. 2002b; Schoenenberger et al. 2005a; Wang et al. 2001; Wang et al. 2000). In all species of the genus, the issue of gene flow is probably largest in Ae. cylindrica due to its pronounced weediness in wheat fields. However, there are at least eleven other Aegilops species that can hybridise acting as female parents with wheat, one of them being Ae. geniculata, which together with Aegilops triuncialis L., is the most widespread species of the genus (Van Slageren 1994). Moreover, hybrids with both bread and durum wheat can be found where the parental species grow together (Van Slageren 1994), for instance in southern Italy and Sicily, but the hybrids' presence is sporadic and inconstant (Pignatti 1982). By planting individual Ae. geniculata plants in durum wheat (Triticum turgidum durum Bowden, 2n=4x=28, AABB genome) stands, David et al. (David et al. 2004) found a hybridisation rate of 0.21%, this frequency was 0.02% if Ae. geniculata seeds were collected from edges of durum wheat fields and germinated. Moreover, chromosome counts of hybrid progeny revealed a high frequency of amphiploids with 56 chromosomes (David et al. 2004).

The aim of the present work was to assess the possibility of gene flow from hexaploid bread wheat to Ae. geniculata under natural field conditions.

Materials and methods

plant material: Aegilops geniculata Roth. seeds originating from four populations (25 seeds per population) (Table 1) were sown in pots in autumn 2002 at the Botanical Garden in Neuchâtel Switzerland, the young rosettes were left over winter for vernalisation. By the end of March 2003 the surviving 53 plants were transplanted in rows in the middle of to two wheat fields in Switzerland. Germination of the plants was reduced because part of the seeds originated from seed exchanges in 1996 and thus were quite old at the moment of sowing. The order of the plants along the rows was randomised by a computer. In order to minimise reproductive interactions, Ae. geniculata individuals were planted at distance of 1 meter from each other. The first wheat field (Field 1) was located in Chatzenrüti at the Swiss Federal Research Station for Agroecology and Agriculture near Zürich (FAL Reckenholz) and cultivated with the Swiss winter wheat variety Zinal. The second (Field 2) was in Fontaines, Val de Ruz, Canton Neuchâtel and cultivated with the Swiss winter wheat variety Arina. Both fields had a surface of approximately 1.5 hectares. In field 1 a total of 23 Ae. geniculata pants were planted, whereas in field two, 29 plants were planted. All seeds of the field grown plants were collected and up to 30 offspring seeds per mother plant were sown at the Botanical Garden under open pollination conditions. Seeds produced by the hybrids in 2004 were collected and sown again. BC1 plants were left to flower, some spikes were covered with pollen-proof bags and others left under open pollination conditions surrounded by a small population of pure Ae. geniculata.

Female fertility was calculated as follows: number of flowers that produced a seed/total number of flowers x 100. All flowers were counted individually.

Table 1. origin of investigated *Ae. geniculata* populations and experimental design

Population	Accession/Location	transplanted	
		Field 1	Field 2
1	403 Giardino Botanico Firenze. Italy, Grosseto, Lago di Burano	12	11
2	Giardino Botanico Siena. Italy Grosseto, Civitella Marittima, ruderal areas, 329 m	4	6
3	Orto Botanico Hanbury Genova. Italy, La Spezia, Devia Marina, 10 m	-	3
4	Croatia, Katici 17° 12' 20" E; 43° 16' 23" N	8	9

Chromosome counts: Root tips for the observation of metaphases were pre-treated in a saturated water solution of α-bromonaphtalene for 180 min at room temperature. Root tips were fixed for at least one week in absolute ethanol and glacial acetic acid (3:1) added with acetocarmine and traces of iron acetate. After that, they were treated in a water solution containing 5% pectinase and 2% cellulase (w/v) for softening of the tissue. The material was then stained in acetocarmine (1%) added with traces of iron acetate for an hour, and heated gently for 2 min over a flame. Root tips were stored in 45% acetic acid, squashed and observed at 1000x magnification under a light microscope.

DNA extraction and PCR: Genomic DNA was extracted from fresh of frozen (-80°C) leaf tissue of a bulk of five individuals of the Swiss wheat varieties Zinal and Arina, from each hybrid and BC1 plant and from Ae. geniculata individuals from the four populations used, using the DNeasy Plant Mini Kit (Qiagen, Basel, Switzerland), following the manufacturer's protocol. Final DNA concentration was adjusted to 20-40 ng/µl, and stored at -20°C. DNA extraction failed in hybrid number 1, because of the bad quality of the collected leaf tissue, however it was included into the statistical analysis due to its clear hybrid morphology. In order to confirm the hybrid nature of the morphologically detected hybrids and BC1s,

PCRs were performed using wheat-specific Sequence Characterised Amplified Regions (SCARs) developed for the purpose of detecting wheat DNA introgressing into Aegilops cylindrica Host (Schoenenberger et al. 2005a). This was the occasion to test whether these markers were useful for detecting introgression into Ae. geniculata too. PCR protocols follow Schoenenberger et al. (2005a).

Results

F1 hybrids: Among the 53 Ae. geniculata plants transplanted into the two wheat fields, one plant originating from population 4 died in field 1 during growth period. The remaining 52 plants flowered at the same time as wheat and produced seeds. A total of 1220 offspring seeds were sown and 850 survived to produce adult plants (69.7%). Among these we detected eight F1 hybrids; overall hybridisation rate was 0.94%. Six hybrids were generated in field 1 whereas the remaining two come from field 2. On a per population base, hybridisation frequencies varied between 0 and 1.94%, whereas if population and field location is taken into account the values were between 0 and 6.06% (Figure 1). The wheat chromosome 6A-specific SCAR DP9 (Schoenenberger et al. 2005a), was detected in all analysed F1 hybrids and absent Ae. geniculata mother populations (Figure 2). Moreover, wheat-specific SCARs D1P9, located on chromosome 6A, and GB10 located on chromosome 5B (Schoenenberger et al. 2005a), were tested on the hybrids and the wild Ae. geniculata populations to see if they were informative in detecting hybridisation between wheat and Ae. geniculata. Marker D1P9 was present in wheat and all hybrids and absent in Ae. geniculata mother populations, whereas GB10 was present in wheat and the hybrids but also in some of the individuals of Ae. geniculata populations 1 and 2 (data not shown).

Eight F1 hybrids produced a total of ten spikes and 92 flowers (Table 2), which carried 2 BC1 seeds. Mean female fertility of the hybrids assessed as the proportion of flowers bearing seeds, was 2.2%, and varied between 0 and 25% per individual.

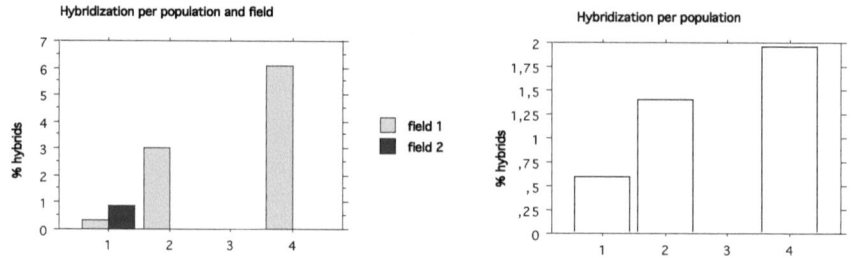

Figure 1. *Ae. geniculata* x *Triticum aestivum* hybridisation frequencies in the field. 1 2 3 4: *Ae. geniculata* populations used for hybridisations.

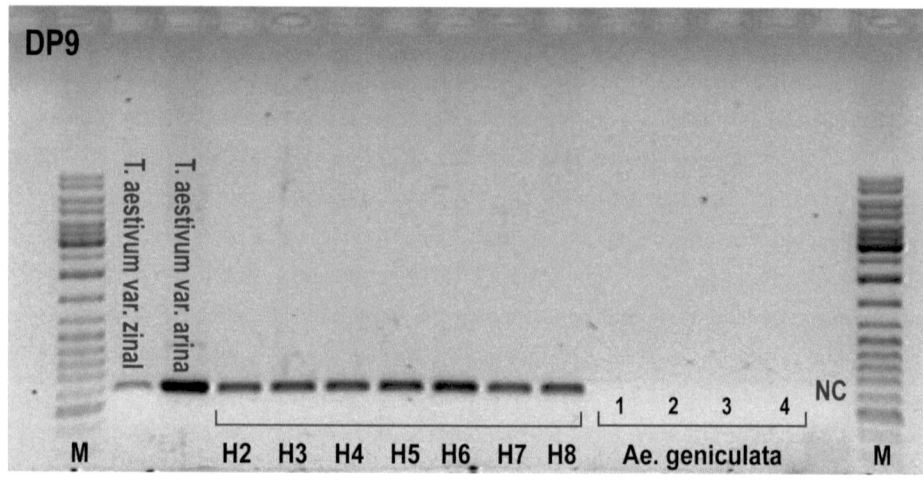

Figure 2. Agarose gel showing wheat chromosome 6A specific SCAR DP9 in the two paternal wheat varieties and in the F1 hybrids to *Ae. geniculata*. The marker is absent in plants from the mother populations 1 to 4. NC negative control, M 100 bp ladder

BC1: Two BC1 seeds were produced one on hybrid 4 and one on hybrid 6. Both their Ae. geniculata progenitors belonged to population 2, and were growing in field 1 where the wheat variety Zinal was planted. Both seeds germinated and produced adult plants. BC1-4 had 56 chromosomes and BC1-6 had 55 chromosomes (Table 2). Both chromosome counts were verified on two distinct metaphase plates. SCARs DP9 and D1P9 were detected in both BC1 plants and absent in their Ae. geniculata progenitor individuals (Figure 3), whereas a test with SCAR IB10 of unknown chromosome location in wheat (Schoenenberger et al. 2005a), resulted in amplification in the Ae. geniculata progenitor individuals of both BC1 plants (data not shown). The two BC1 plants did not produce any offspring, a total of 53 spikes produced 1076 florets but 0 seeds.

Table 2. Fertility of F1 *Ae. geniculata* x *Triticum aestivum* F1 hybrids and chromosome numbers of BC1s

Hybrid/BC1	Spikes	flowers	fertility	BC1 seeds	Chromosome Number BC1
1	1	4			
2	3	32			
3	1	12			
4	1	14	7.14%	1	56
5	1	14			
6	1	4	25%	1	55
7	1	8			
8	1	4			
total/mean	10	92	2,17%	2	

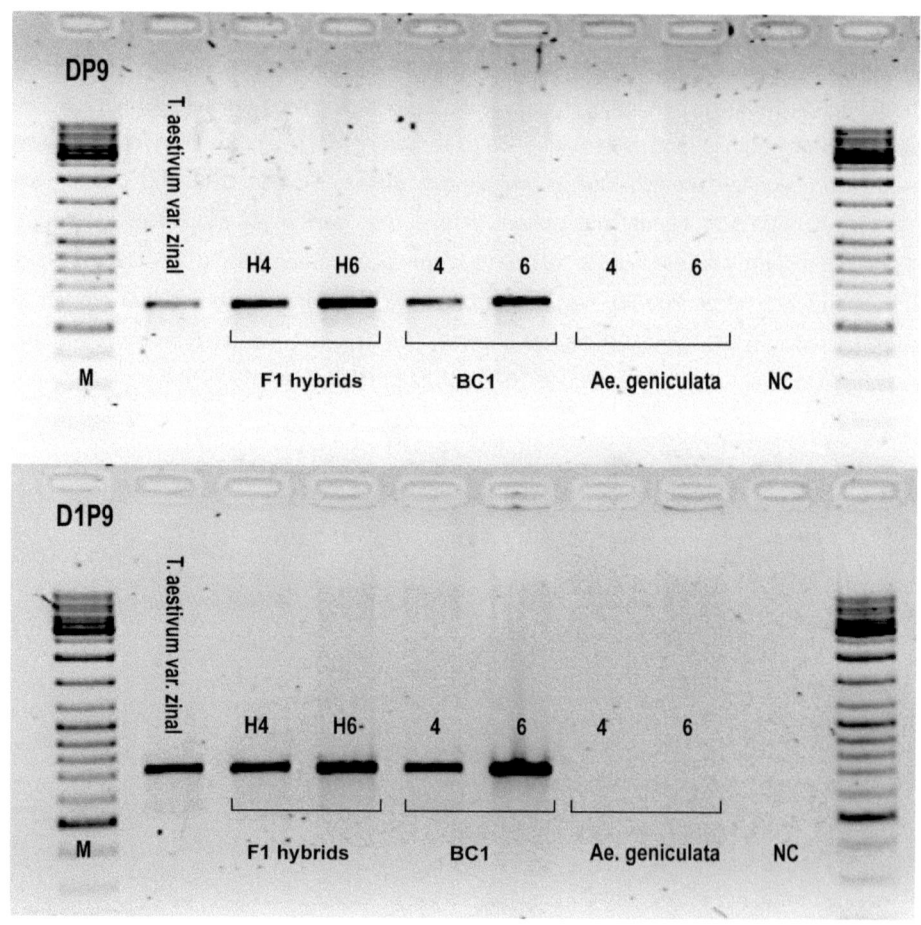

Figure 3. Agarose gel showing SCARs DP9 (specific to wheat chromosome 6A) and D1P9 (specific to wheat chromosome 6D) in the paternal wheat variety, in F1 hybrids and their BC1 offspring, absent in the *Ae. geniculata* maternal progenitor individuals. NC negative control, M 100 bp ladder

Discussion

Overall Ae. geniculata x Triticum aestivum hybridisation rate was 1.7% in field 1 and 0.4% in field 2, but if Ae. geniculata populations and field locations are taken into account the rate could be as high as 6% (Figure 1). These frequencies are higher than the observed Ae. geniculata x Triticum turgidum durum hybridisation rates of 0.21% in nursery conditions, and 0.02% at agricultural scale field edges (David et al. 2004). Indeed, experimental designs differed between both studies as we planted Ae. geniculata in the middle of agricultural scale wheat fields, whereas in nursery conditions, the wheat pollen pressure was surely much smaller. Moreover, wheat pollen that may be competitive in fertilising another species does not fly far from a field. In fact, intraspecific gene flow from bread to durum wheat decreased from 0.04% to 0.02% and from 0.03% to 0.01% in two different years between 1 and 5 meters from the pollen source (Matus-Cadiz et al. 2004). Moreover, hybrids between Ae. cylindrica and wheat were detected only up to 1m of the wheat field (0.29%; Schoenenberger et al. 2005b).

Our Ae. geniculata x T. aestivum hybridisation rates between 0 and 6% established in the field, are similar to the ones found for Ae. geniculata and Triticum turgidum durum (David et al. 2004). They are also of the same order of magnitude than those observed for Ae. cylindrica x T. aestivum hybridisation which varied between 0 and 8% on a per field basis in USA (Morrison et al. 2002b), and between 1 and 7% depending on the Ae. cylindrica population in Switzerland (Guadagnuolo et al. 2001a). Interestingly, hybrid fertility was higher in Ae. geniculata hybrids (2.2%) than in Ae. cylindrica hybrids where it was 1% (Morrison et al. 2002a),. This may indicate an increased exposure of Ae. geniculata to gene flow from wheat. In fact, some Ae. geniculata x T. aestivum hybrids display a high frequency of homeologous pairing, demonstrating relatedness between Ae. geniculata and T. aestivum genomes (Fernandez-Calvin and Orellana 1992). Moreover, agronomical interesting genes have been transferred from Ae. geniculata to T. aestivum by homeologous recombination, and disomic addition lines have been generated e.g. (Aghaee-Sarbarzeh et al. 2002), indicating that the wheat ABD genomes and the Ae. geniculata UM genomes are homeologous enough to allow gene flow to occur.

Analysis of SCARs which were developed for the purpose of tracing wheat-specific DNA in introgressed Ae. cylindrica plants (Schoenenberger et al. 2005a) has shown to be useful for introgression studies on Ae. geniculata too. However, not all the SCARs were wheat-specific when compared to Ae. geniculata populations. In fact, out of four amplified markers, only two were wheat-specific, and the other two were amplified in some of the wild individuals too. It is difficult though to state whether a marker being constantly present in wheat and sporadic in the wild species has a history of common ancestry or if it ended up in the wild genome due to gene flow occurred in the past. This is particularly true for the accessions investigated here which partially come from seed exchanges among botanical gardens and it is almost impossible to know if they grew in with wheat at some point of their history.

The BC1 individuals produced had 56 and 55 chromosomes which are the same values as observed in Ae. geniculata x Triticum turgidum durum amphiploid backcrosses (David et al. 2004). However, durum wheat is a tetraploid with 28 chromosomes whereas bread wheat has 42. In Ae. cylindrica x T. aestivum backcrosses, chromosome numbers often were around 40-42 and never 56 (Schoenenberger et al. 2005c). To get to a value of 56 one must imagine either a female gamete of 42, fecundated by a male gamete of 14 chromosomes, or a female gamete of 28, fertilised by an unreduced Ae. geniculata pollen grain with 28 chromosomes. However these BC1 plants were completely sterile either selfed of left under open pollination conditions. Increased chromosome load could be an explanation for this observation, but it seems more likely that the pollen pressure of the few surrounding Ae. geniculata plants was not enough to fertilise the BC1s, even more if we take into account the fact that the BC1s were much bigger in size, and their spikes placed well above the Ae. geniculata inflorescences. To conclude we can say that gene flow from bread wheat to Ae. geniculata is likely to occur where the two species growing sympatry, and its frequency may even be higher than to Ae. cylindrica.

Acknowledgements

We are grateful to Franz Bigler of the Swiss Federal Research Station for Agroecology and Agriculture in Zürich Reckenholz for allowing the field experiments, to Anouk Béguin and Mei Lin Cheung for technical assistance, and to the team of the Botanical Garden of Neuchâtel for its help in the cultivation of the plants. Finally, we would like to thank Nils Arrigo for critical discussion and relecture of the manuscript and to Philippe Küpfer for his support. This project was funded by the National Centre of Competence in Research (NCCR) Plant Survival, a research programme of the Swiss National Science Foundation.

References

Aghaee-Sarbarzeh, M., M. Ferrahi, S. Singh, H. Singh, B. Friebe, B. S. Gill, and H. S. Dhaliwal. 2002. Ph-I-induced transfer of leaf and stripe rust-resistance genes from Aegilops triuncialis and Ae. geniculata to bread wheat. Euphytica. 127:377-382.

David, J. L., E. Benavente, C. Brès-Patry, J.-C. Dusautoir, and M. Echaide. 2004. Are neopolyploids a likely route for a transgene walk to the wild? The *Aegilops ovata* x *Triticum turgidum durum* case. Biological Journal of the Linnean Sociey. 82:503-510.

Ellstrand, N. C. 2003. Dangerous Liaisons? When Cultivated Plants Mate with Their Wild Relatives. Baltimore, The Johns Hopkins University Press.

Fernandez-calvin, B. and J. Orellana. 1992. Relationship between Pairing Frequencies and Genome Affinity Estimations in Aegilops-Ovata X Triticum-Aestivum Hybrid Plants. Heredity. 68:165-172.

Guadagnuolo, R., D. Savova-Bianchi, and F. Felber. 2001a. Gene flow from wheat (*Triticum aestivum* L.) to jointed goatgrass (*Aegilops cylindrica* Host.), as revealed by RAPD and microsatellite markers. Theoretical and Applied Genetics. 103:1-8.

Matus-Cadiz, M. A., P. Hucl, M. J. Horak, and L. K. Blomquist. 2004. Gene flow in wheat at the field scale. Crop Science. 44:718-727.

Morrison, L. A., L. Crémieux, and C. A. Mallory-Smith. 2002a. Infestations of jointed goatgrass (*Aegilops cylindrica*) and its hybrids in Oregon wheat fields. Weed Science. 50:737-747.

Morrison, L. A., O. Riera-Lizarazu, L. Crémieux, and C. A. Mallory-Smith. 2002b. Jointed Goatgrass (*Aegilops cylindrica* Host) X Wheat (*Triticum aestivum* L.) Hybrids: Hybridization Dynamics in Oregon Wheat Fields. crop Science. 42:1863-1872.

Pignatti, S. 1982. Flora d'Italia. Bologna.

Schoenenberger, N., F. Felber, D. Savova-Bianchi, and R. Guadagnuolo. 2005a. Introgression of wheat DNA markers from A, B and D genomes in early generation progeny of *Aegilops cylindrica* Host X *Triticum aestivum* L. hybrids. Theoretical and Applied Genetics. DOI: 10.1007/s00122-005-0063-7:1-9.

Schoenenberger N., D. Savova-Bianchi, R.Guadagnuolo, and F. Felber. 2005b. Gene flow by pollen from wheat to *Aegilops cylindrica* Host as a function of distance: a field experiment. *submitted*

Schoenenberger N., D. Savova-Bianchi, R.Guadagnuolo, and F. Felber. 2005c. Introgression from transgenic wheat into *Aegilops cylindrica* Host: molecular analysis, cytogenetics and fertility of hybrids, BC1 and BC1S1. *Submitted*

Van Slageren, M. W. 1994. Wild wheats: a monograph of *Aegilops* L. and *Amblyopyrum* (Jaub. & Spach) Eig (Poaceae), Wageningen Agricultural University Press, Wageningen; ICARDA, Aleppo.

Wang, Z., R. S. Zemetra, J. Hansen, and C. A. Mallory-Smith. 2001. The fertility of wheat x jointed goatgrass hybrid and its backcross progenies. Weed Science. 49:340-345.

Wang, Z. N., A. Hang, J. Hansen, C. Burton, C. A. Mallory-Smith, and R. S. Zemetra. 2000. Visualization of A- and B-genome chromosomes in wheat (Triticum aestivum L.) x jointed goatgrass (Aegilops cylindrica Host) backcross progenies. Genome. 43:1038-44.

Wolfenbarger, L. L. and P. R. Phifer. 2000. Biotechnology and ecology - The ecological risks and benefits of genetically engineered plants. Science. 290:2088-2093.

Die VDM Verlagsservicegesellschaft sucht für wissenschaftliche Verlage abgeschlossene und herausragende

Dissertationen, Habilitationen, Diplomarbeiten, Master Theses, Magisterarbeiten usw.

für die kostenlose Publikation als Fachbuch.

Sie verfügen über eine Arbeit, die hohen inhaltlichen und formalen Ansprüchen genügt, und haben Interesse an einer honorarvergüteten Publikation?

Dann senden Sie bitte erste Informationen über sich und Ihre Arbeit per Email an *info@vdm-vsg.de*.

Sie erhalten kurzfristig unser Feedback!

VDM Verlagsservicegesellschaft mbH
Dudweiler Landstr. 99   Telefon +49 681 3720 174
D - 66123 Saarbrücken   Fax   +49 681 3720 1749
www.vdm-vsg.de

Die VDM Verlagsservicegesellschaft mbH vertritt

Printed by Books on Demand GmbH, Norderstedt / Germany